UF

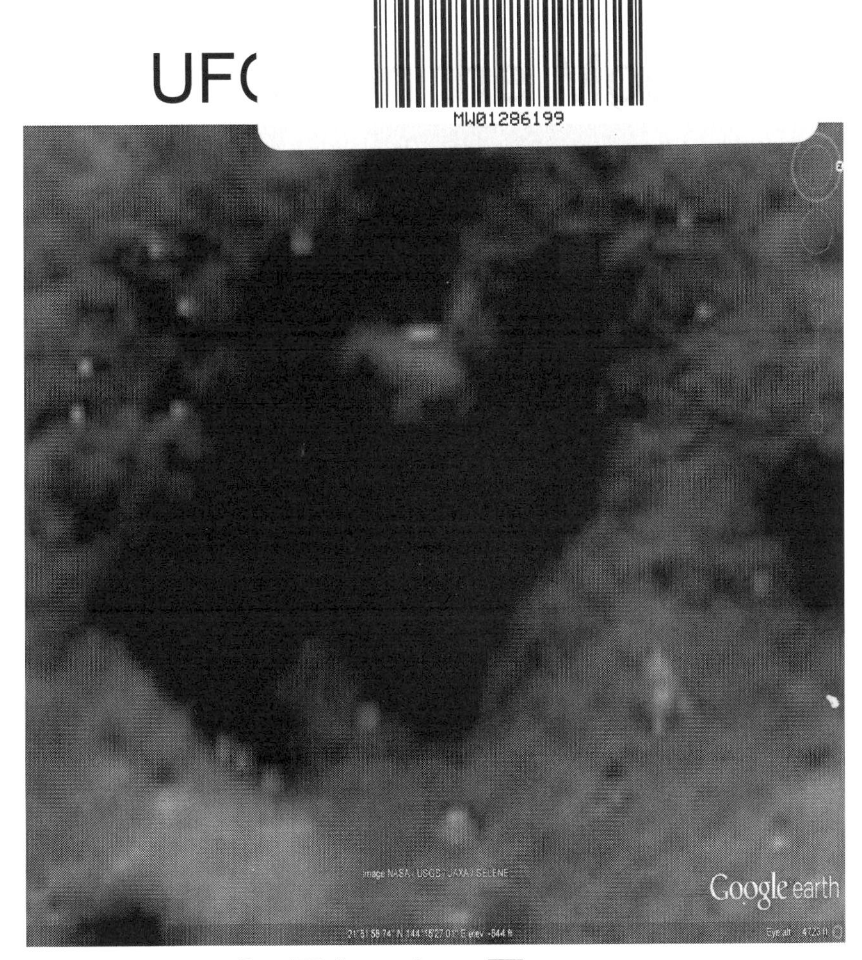

A Photo Essay

By

Kenneth Clark

ISBN:1477448217
ISBN-13:978-1477448212

DEDICATION

In 1996 I decided to study and research UFOs, and write a book about UFOs so everybody would know what I knew about UFOs. What I had found out changed my life. This is very important to me and soon you will see why it should be important to you.

I use the Internet for my research and I couldn't believe all the things I found about UFOs. Then I heard about Google Earth, and I downloaded it. Google Earth is a miracle, I could not believe all the things I found on it. Then I tried Google Mars , unbelievable beautiful pictures of craters and valleys , I loved it. Every day I would look on Mars to see what I could find. Then one day I saw a straight white line, which is not natural on Mars as I zoomed down on the white line I noticed that it was in the bottom of a valley, it went off the cliff side disappearing through sand dunes and coming out the other side, this was a white pipe that went up... across valleys through craters in a perfectly straight line. I knew it wasn't natural so I zoomed in on it and studied it very closely and found out that it was a pipe made by somebody. I looked for more stuff in that close up to the ground view of Mars, and a half-mile to the right of the pipe, I found another straight white pipe. Both pipes went through the entire picture, they were 11 miles long. These pipes are engineering marvels, I had stumbled upon a massive public works project on the surface of Mars. The more I looked , the more I discovered. It is all right there for anyone to see. It was at that point that I decided to investigate the moon in the same way.

I had heard about Google moon. So I decided to look for lights on the moon. With in two minutes I found a group of objects in a crater on the surface of the moon. I zoomed in and adjusted the light setting and brightened the picture up. I was amazed I was looking at a long cigar shaped UFO several dots and a square object and many more dots in the background. These weren't pipes or buildings they were spacecraft!

I looked for more within seconds a huge group of oddly shaped objects, so I took another picture on my computer, I spent all afternoon and found dozens of objects and took pictures of all of it and saved it to my files on my computer. The next day I spent hours taking pictures and documenting all that I had found. Some of the C class or cigar shaped UFOs I measured to be about 200 feet, two were 1000 feet long. The USS Enterprise Aircraft Carrier is 1123 feet long by comparison. Two spacecraft as big as an aircraft carrier!

After a month of taking hundreds of UFO pictures and filing them away on my computer, I found my first V class UFO. How I had overlooked these earlier I don't know they were right in front of me. These are giant V-shaped UFOs, they had 4 to 600 feet wingspan. I found smaller V class ships , and some very large ones. I now have 392 pictures of V class UFOs on file.

I also found huge cities on the moon. Lots of tall towers with lights on top of them. There are UFOs and structures all over the moon. The focus patch which is 200 miles **square**, is on the dark side of the Moon. All the UFOs are parked on the shadows edge, around craters. There must be a huge underground city on the moon, all over the moon. I found one UFO which was round in a crater, almost a mile across. Bright Silver with a dot in the middle. I couldn't believe it all these

UFOs on the moon. They are not on the Internet anywhere. How can I be the only one that knows about them?

I contacted all the major media outlets and got the cold shoulder. No interest. It is really amazing how you can stand there with a picture in your hand and have someone deny that UFOs exist.

I contacted Colleges and Universities and received the same response. I tried to contact astronomers to get some confirmation and just to make sure I was interpreting the images correctly and they refused to examine them. I would think that a scientist, in search for the truth, could spend a couple of minutes looking at a photo but they ran in the other direction. I find that a little odd, don't you?

I have sent hundreds of requests to NASA for confirmation and interpretation of the pictures, many are sourced from NASA, and they ignore my requests. I am not a big fan of conspiracy theories but all this could make me a believer.

Let me make this clear, I don't want people to panic or get worried that aliens are our next door neighbors. These structures and ships are massive and didn't appear overnight. These aliens have been on the moon, mars and probably hidden on earth for decades and perhaps centuries. If they wanted to hurt us they could of easily wiped us out with their technology. They obviously want to observe else and hopefully help us.

The release of these photos by NASA and others indicates that something big is about to happen. The various governments of the world have kept aliens under wraps for a very long time and could continue to do so but I think that something has changed and they are letting this information leak out bit by bit so that when the official announcement comes there will be acceptance of our 'new' friends.

One of the things that I believe has changed is the impending approach of 'Planet X' in December of 2012 which was predicted in the Mayan Calendar many years ago. When 'Planet X' comes, it combined with a unique planetary alignment will cause a great flood worldwide. That combined with extreme solar flares caused by the gravitation impact of 'Planet X' indicate a possible extinction event. Possibly the only way to save humanity is with outside help.

NASA has put out an official warning concerning extreme solar flares for December 2012. Planet X will be here in December 2012. It is time for the truth to be revealed.

Well that's how I found the UFOs on the moon and that is a very small part of my story. I just think it is time now to show everybody all over the world these pictures of the UFOs over the moon why wait? Everybody needs to know and wants to know. Were all in the same boat together and we need to move into the future as soon as possible and leave the evil, money mad, world behind. . Thanks for your consideration.

Sincerely,

Kenneth Clark

CONTENTS

ACKNOWLEDGMENTS

Many thanks to Richard Booher for his assistance in compiling and editing this book.

The photos are from Google Earth and are resourced from NASA, JPL USGS, JAXA, SELENE, and other public domain resources.

1 JOURNEY TO THE MOON

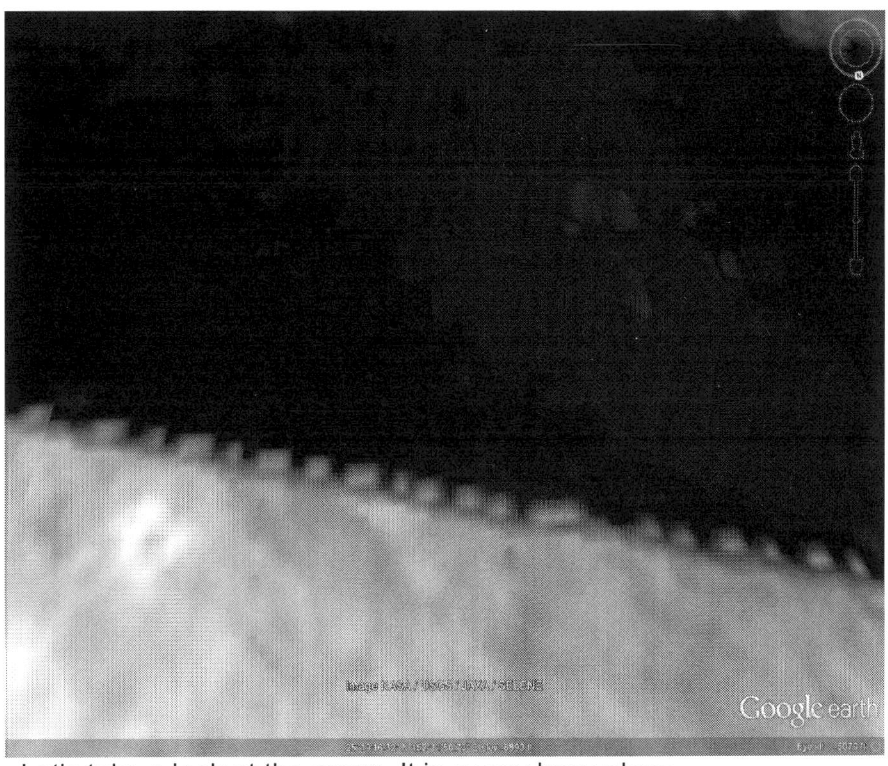

Let's take a look at the moon. It is a very busy place.

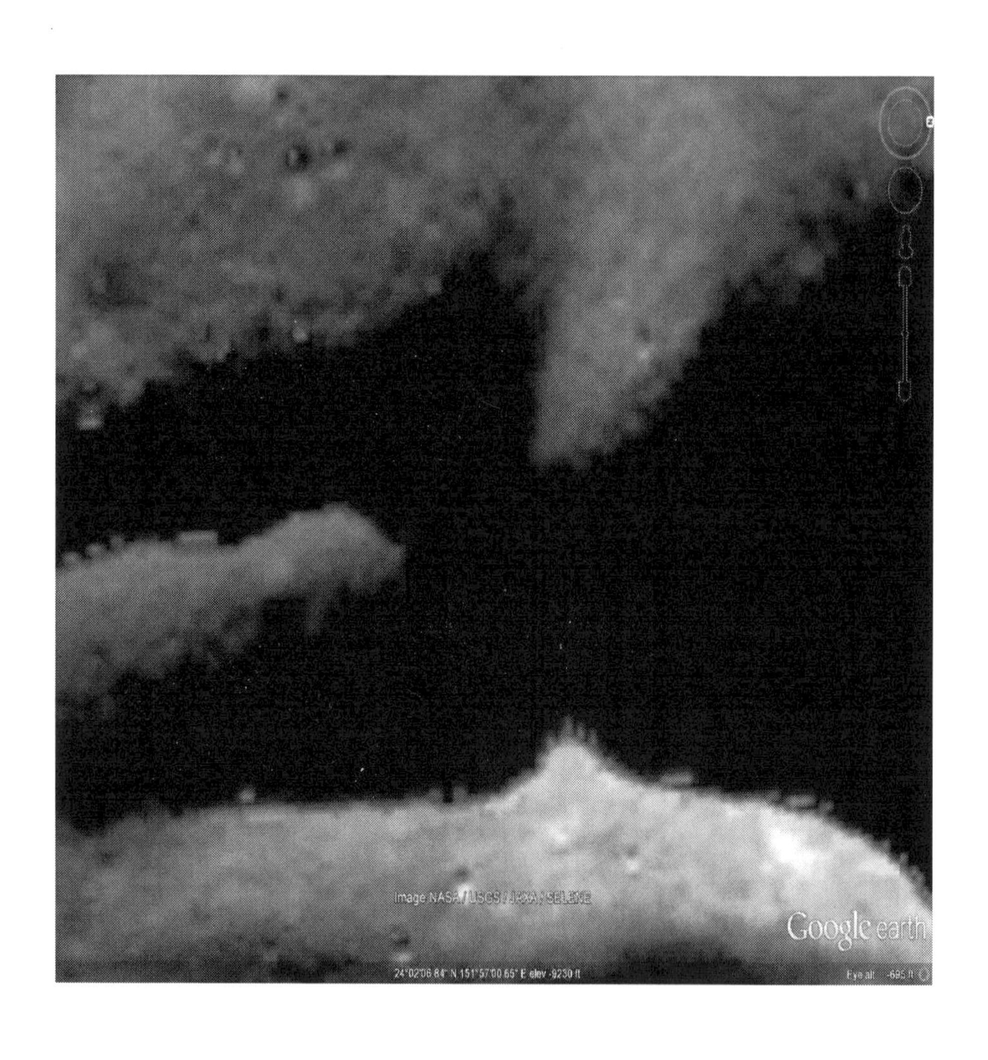
Image NASA / USGS / JAXA / SELENE
Google earth
24°02'06.84" N 151°57'00.65" E elev -9230 ft
Eye alt -695 ft

Two large V spacecraft in craters.

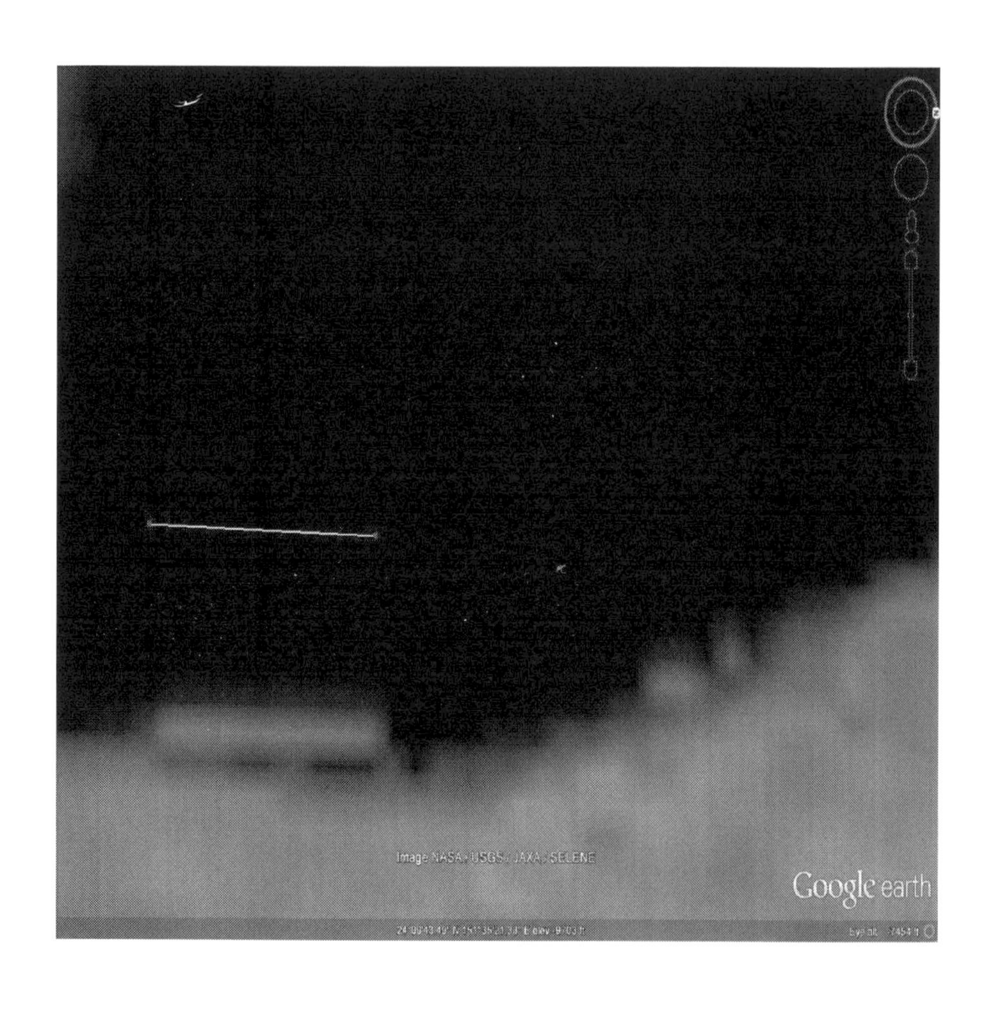

600 foot cigar shaped UFO

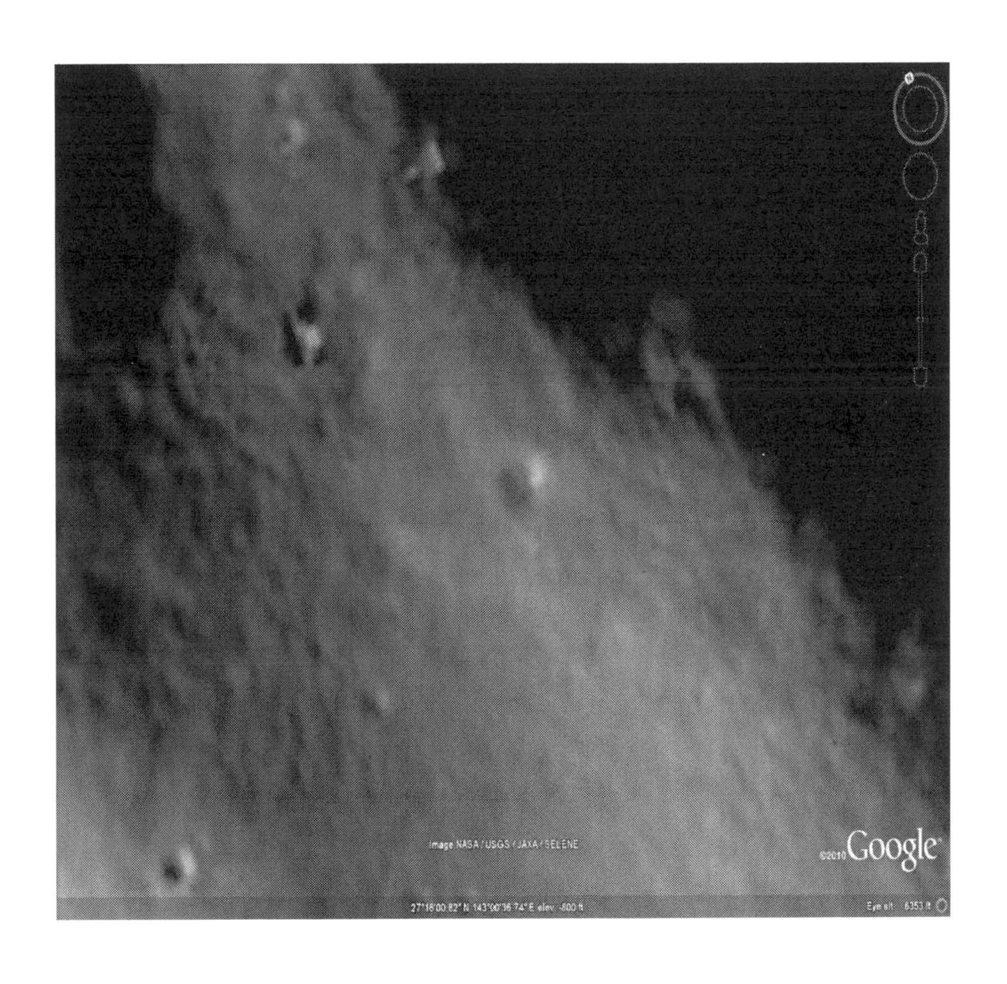

UFO's and Buildings on Craters Edge

Largest UFO found on the moon.

UFO's and Buildings on craters edge.

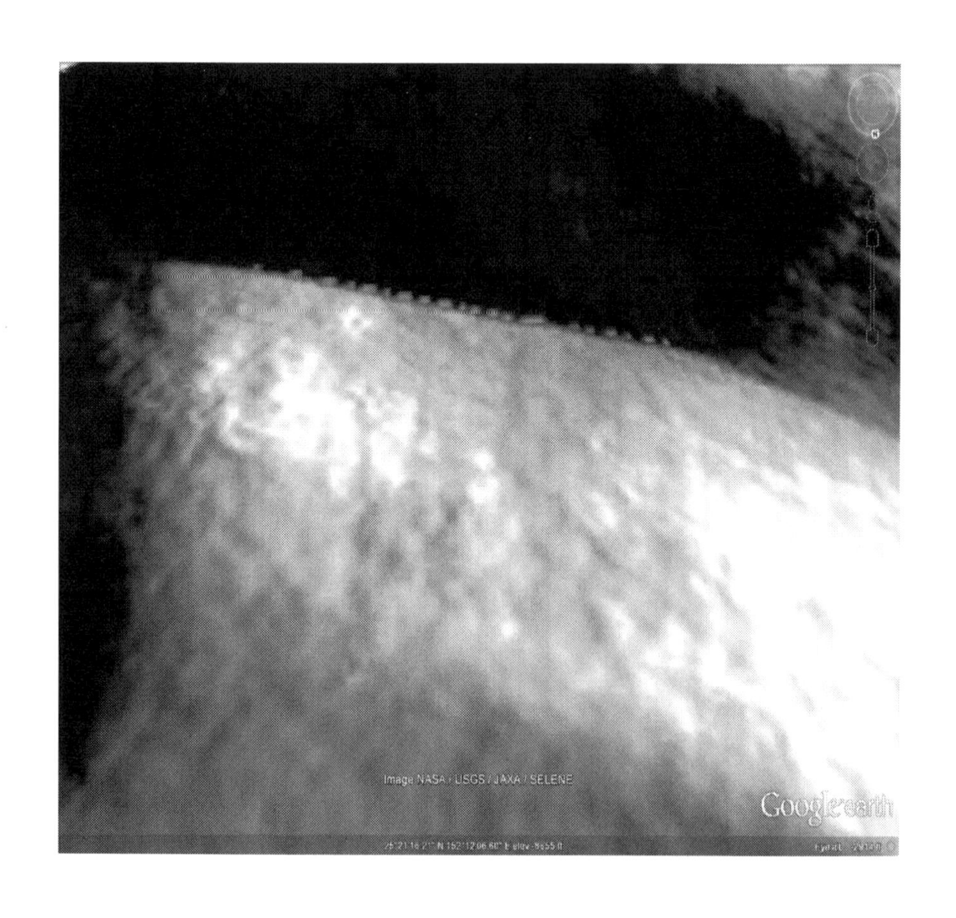

More objects on craters edge

Image NASA / USGS / JAXA / SELENE
Google
30°10'35.84" N 142°48'21.55" E elev -4046 ft

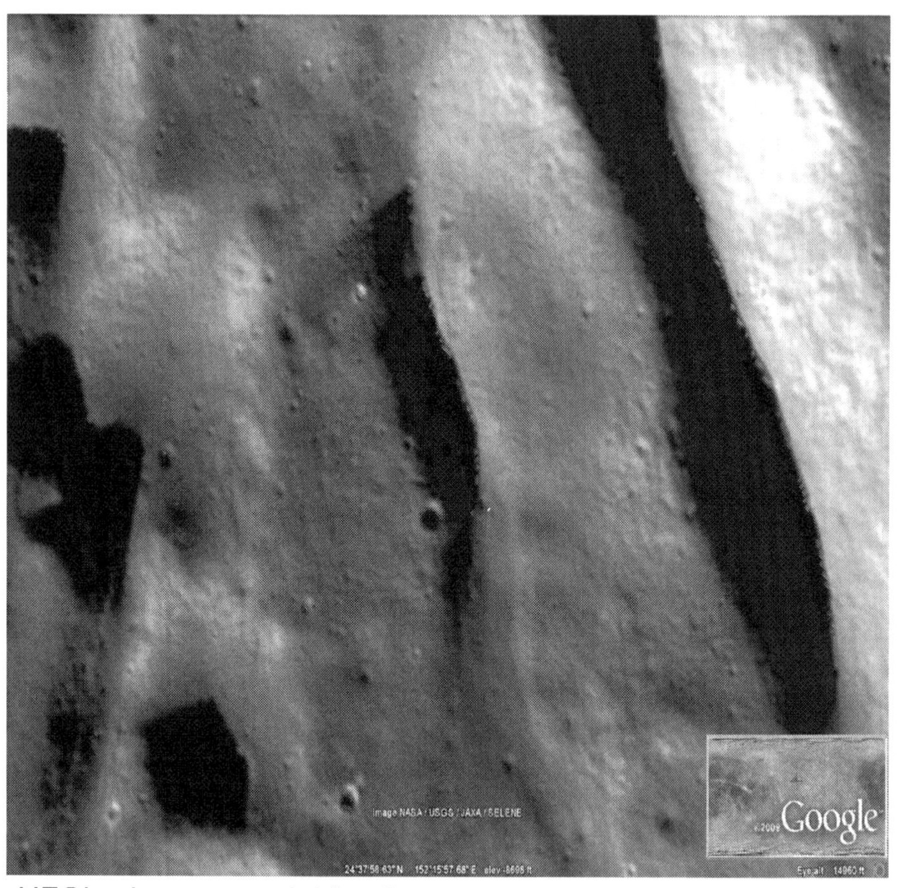

UFO's along several ridge lines

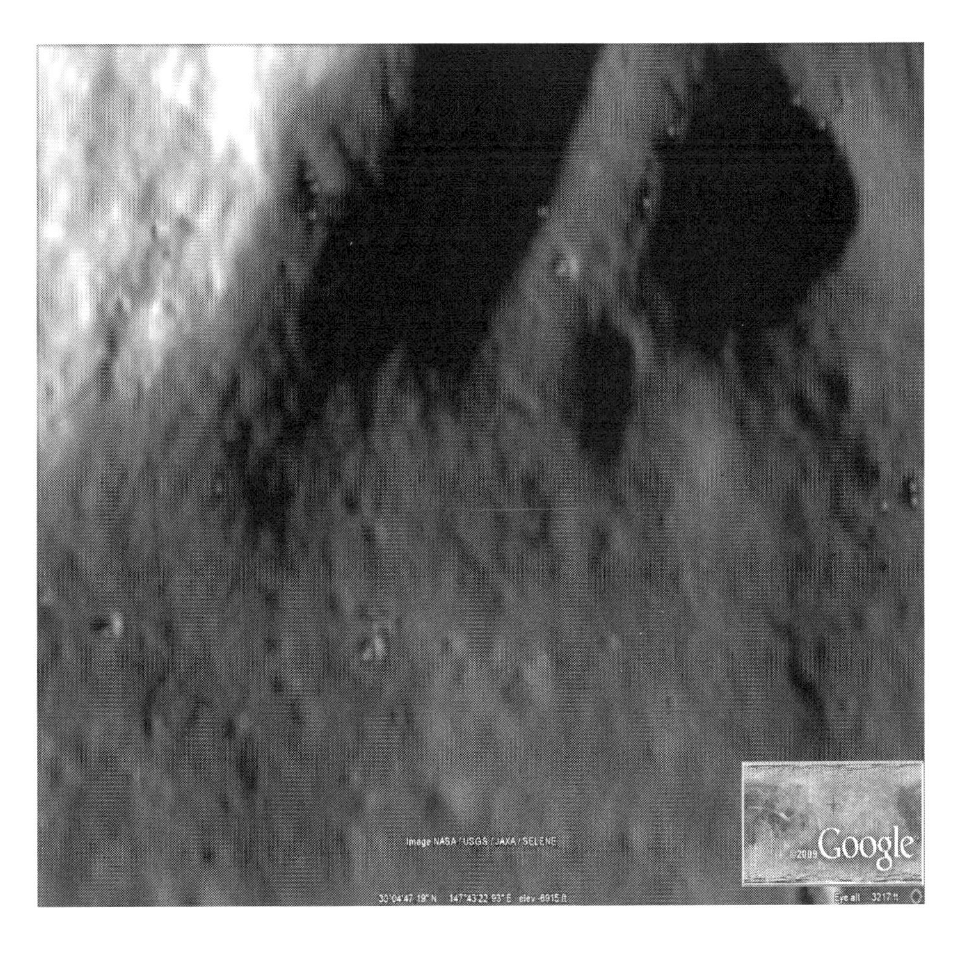

Valley of UFO's

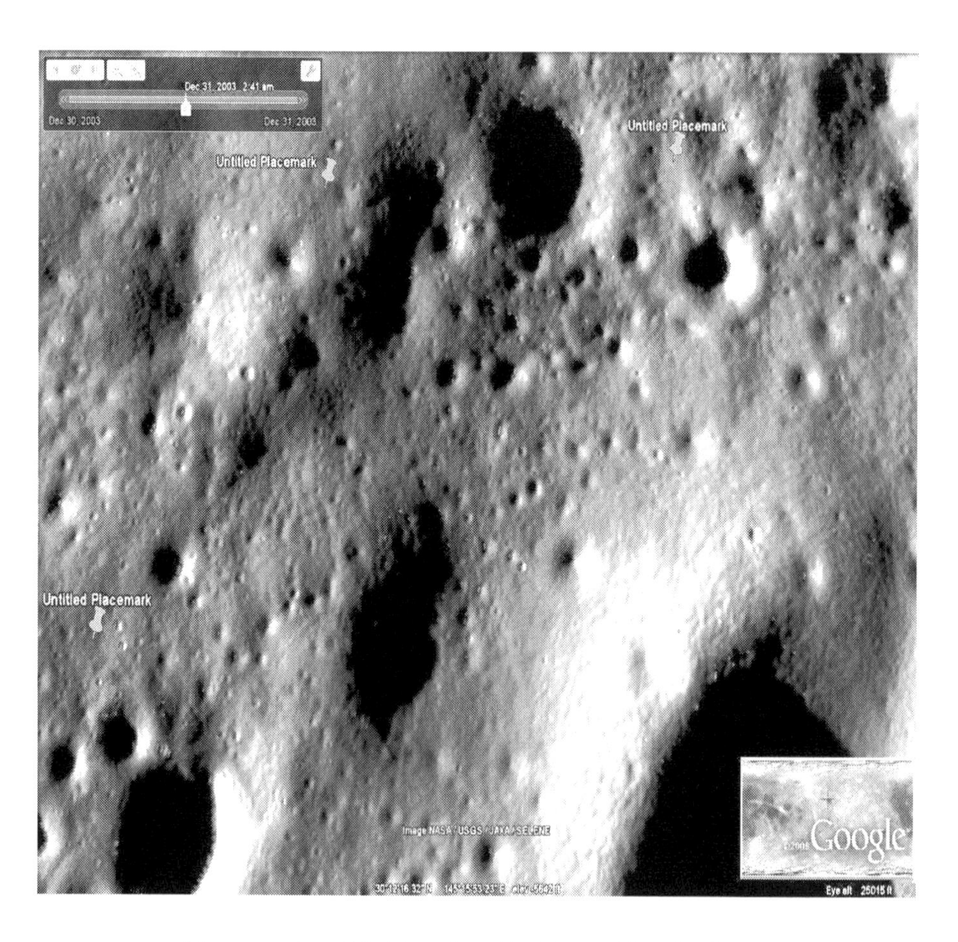

More parked UFO in the Northern Hemisphere

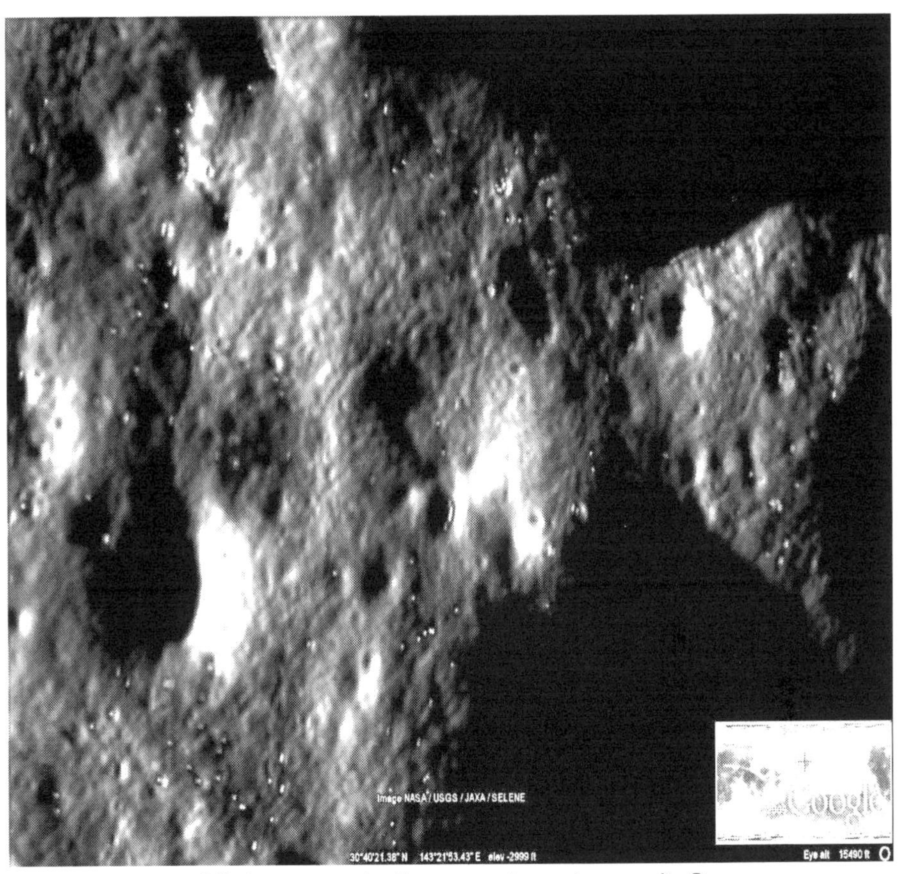

High concentration, maybe a large city?

Valley of coils (each ¼ mile in length)

At home on the moon

Large cigar UFO top right crater

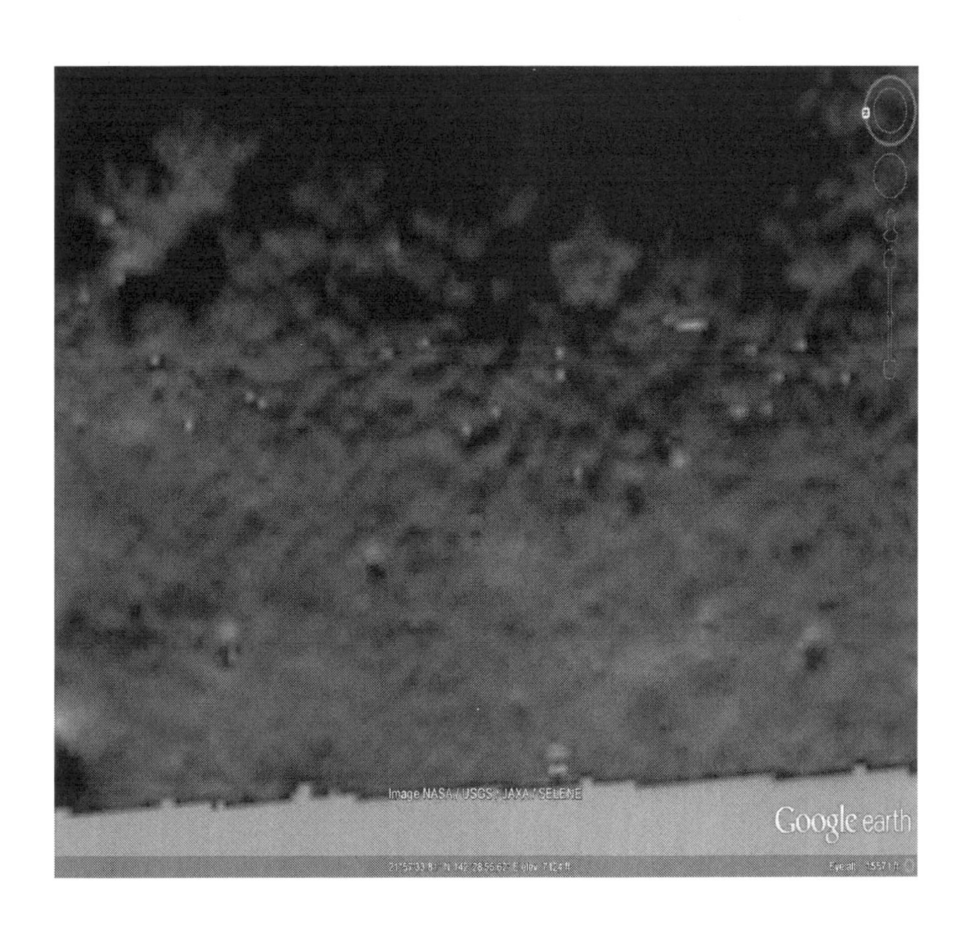
Image NASA / USGS / JAXA / SELENE
Google earth

Image NASA / USGS / JAXA / SELENE
21°48'45.77" N 143°12'55.23" E elev 1923 ft
©2009 Google
Eye alt 20933 ft

"U" shaped UFO

UFO Cluster

Huge city with recognizable buildings

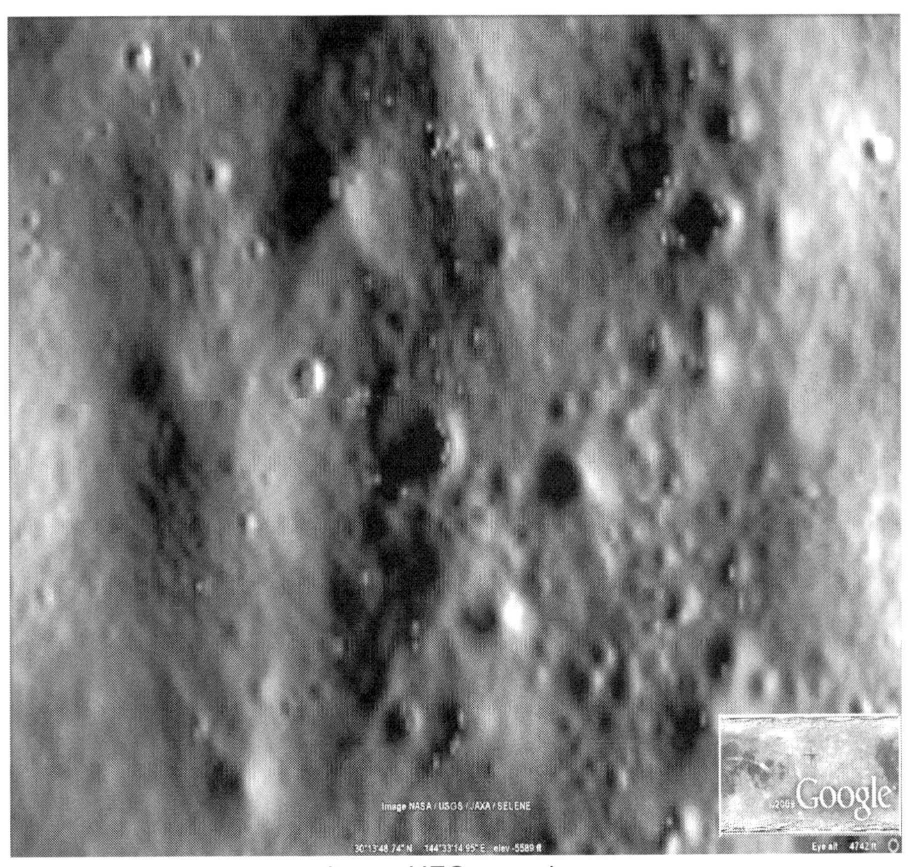

Large UFO grouping

Cities with recognizable buildings

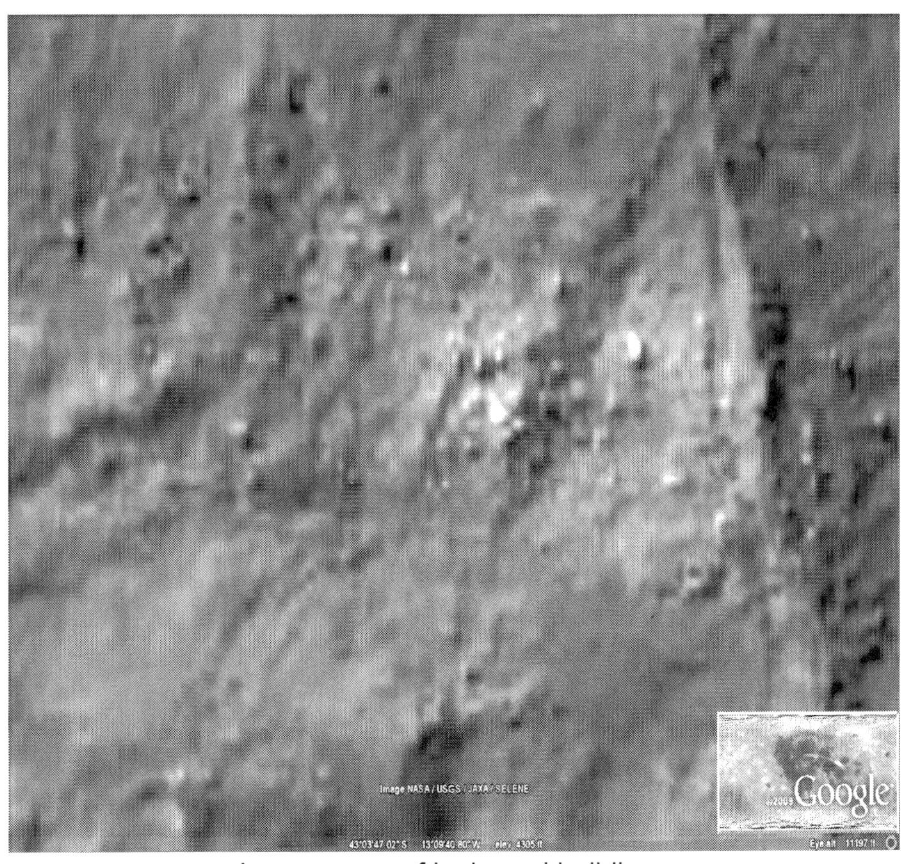

Large group of L shaped buildings

UFO's on shadows edge

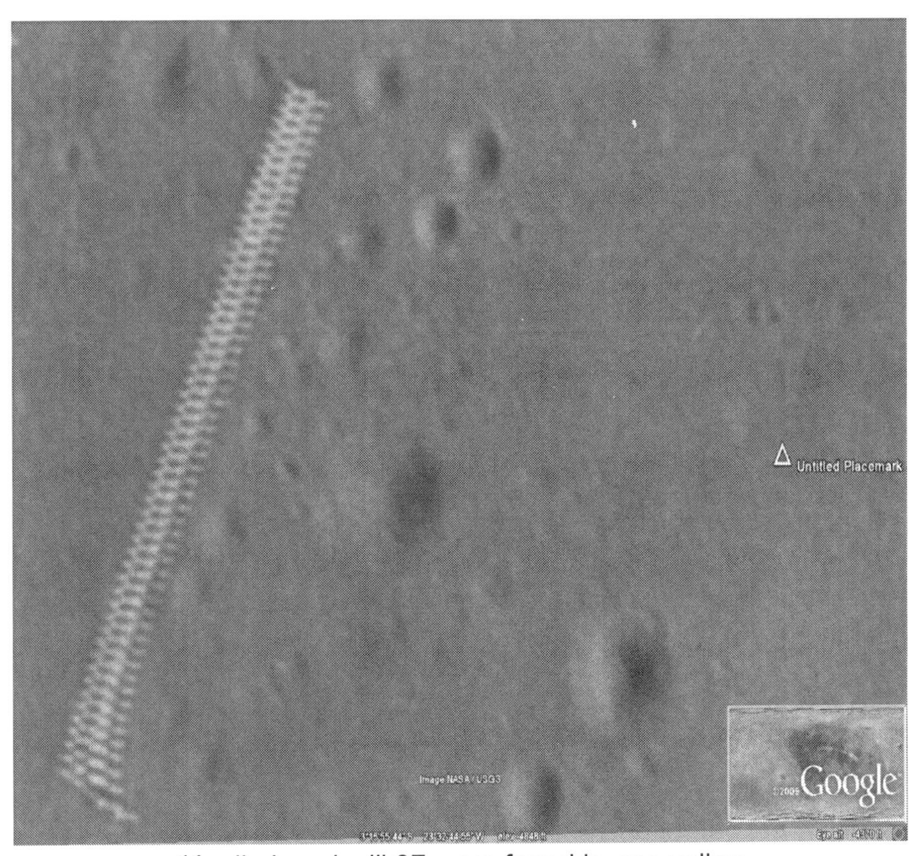

¼ mile long 'coil' 37 were found in one valley

UFO's reflecting bright lights

High contrast view of parking area

Several different types of UFO's parked

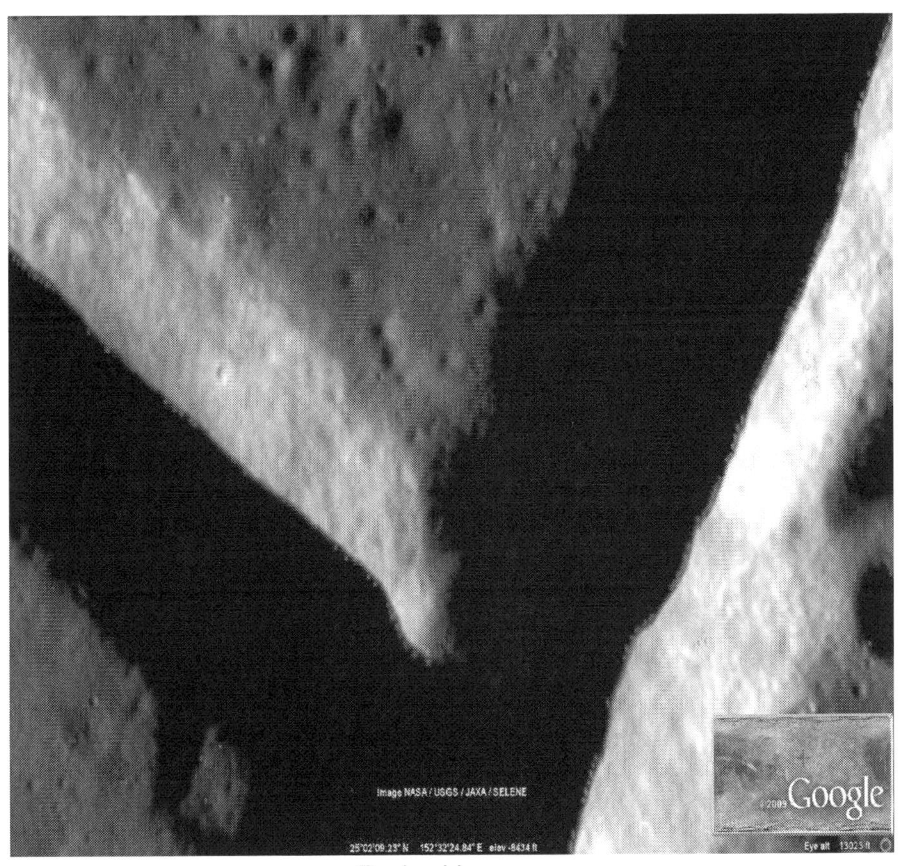

Parked in rows

Small city on cliff

UFO's parked near 'bottomless' crater

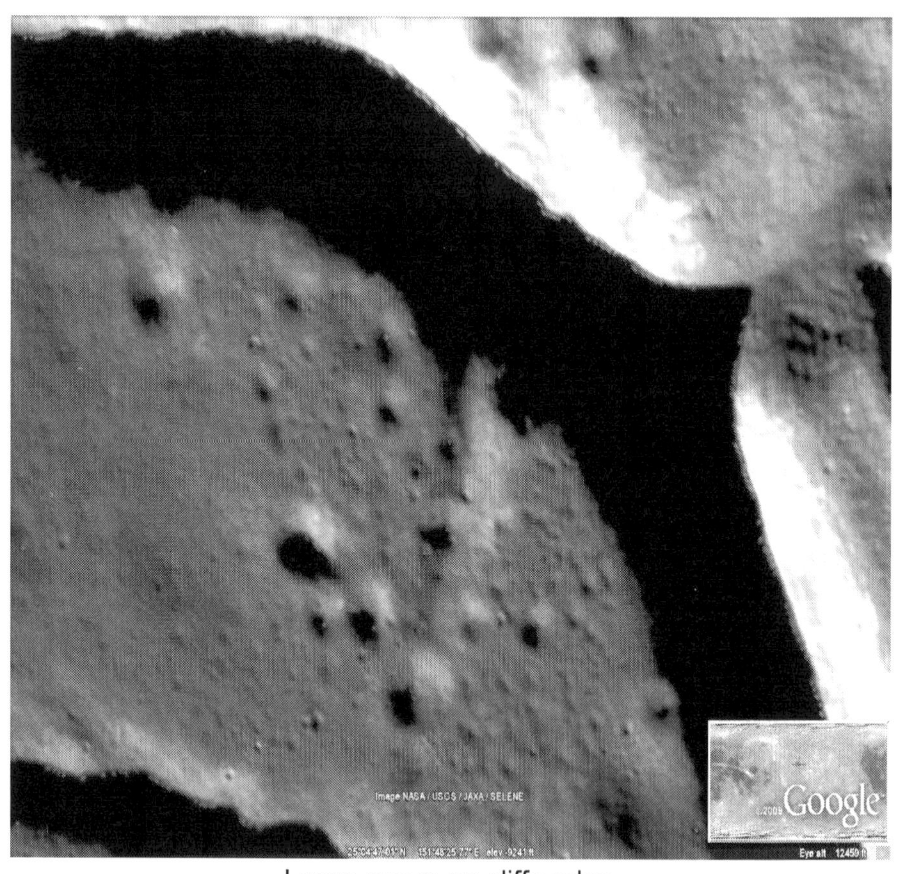

Large group on cliffs edge

A variety of cigar UFOs of different lengths

Sometimes it seems like there is a UFO in every crater!

Image NASA / USGS / JAXA / SELENE
30°52'55.73" N 144°31'36.33" E elev -5641 ft
Google
Eye alt 17456 ft

All types of buildings and UFOs

T shaped building in center

900 foot cigar UFO on right

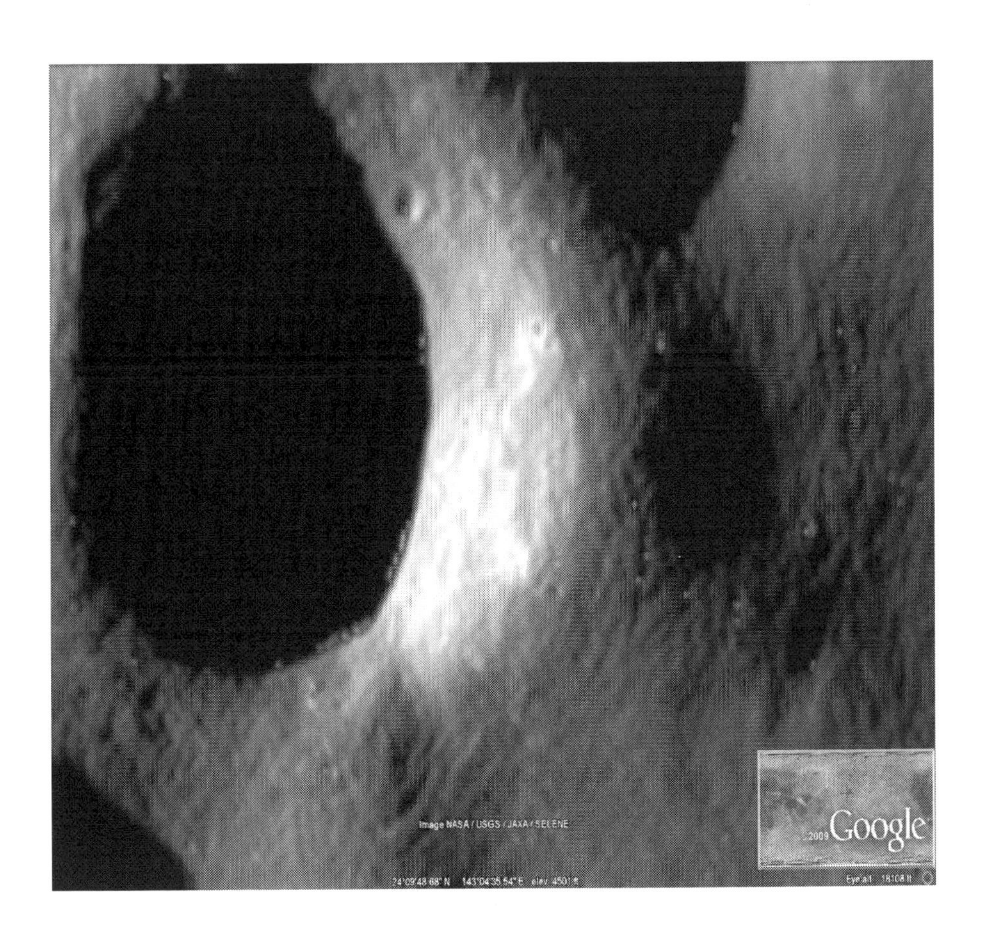
Image NASA / USGS / JAXA / SELENE
©2009 Google
24°09'48.68" N 143°04'35.54" E elev 4501 ft
Eye alt 18108 ft

3 'V' UFOs across the picture

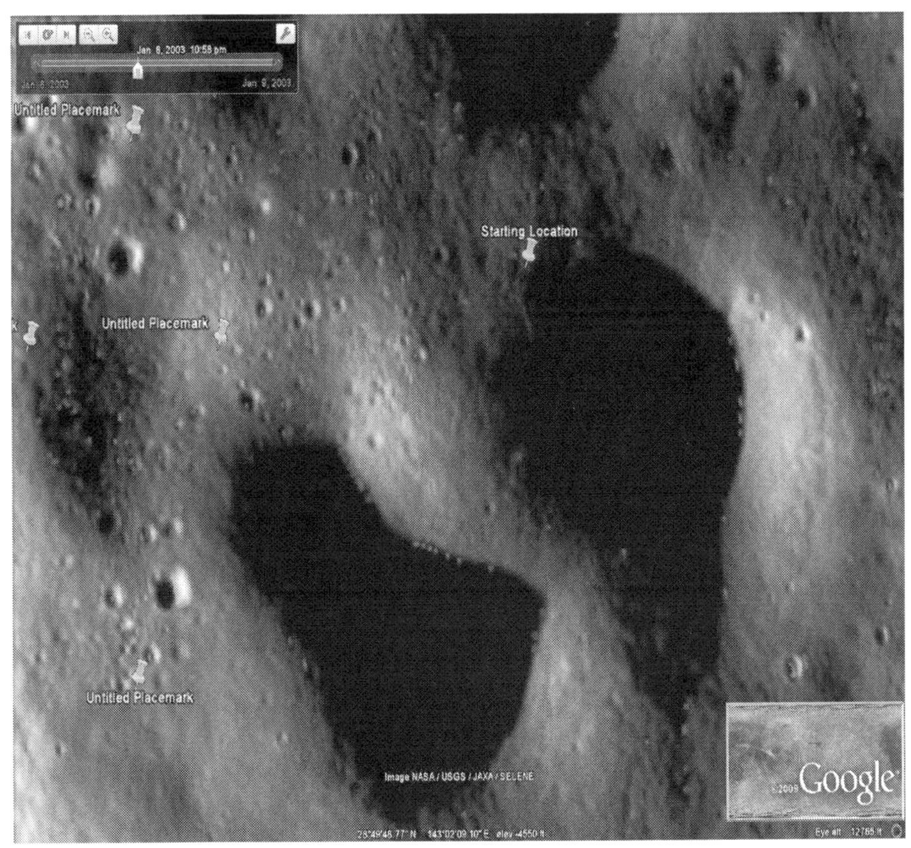

Small grouping of UFOs on left

'V' ship on right corner

Image NASA / USGS / JAXA / SELENE
Google earth

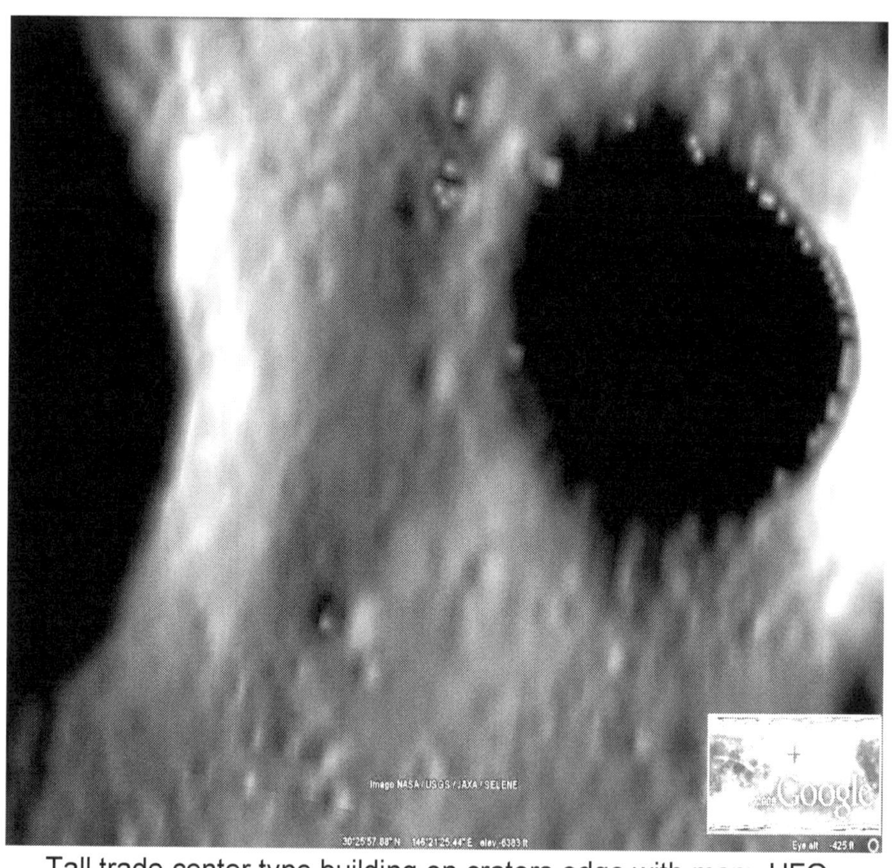

Tall trade center type building on craters edge with many UFOs

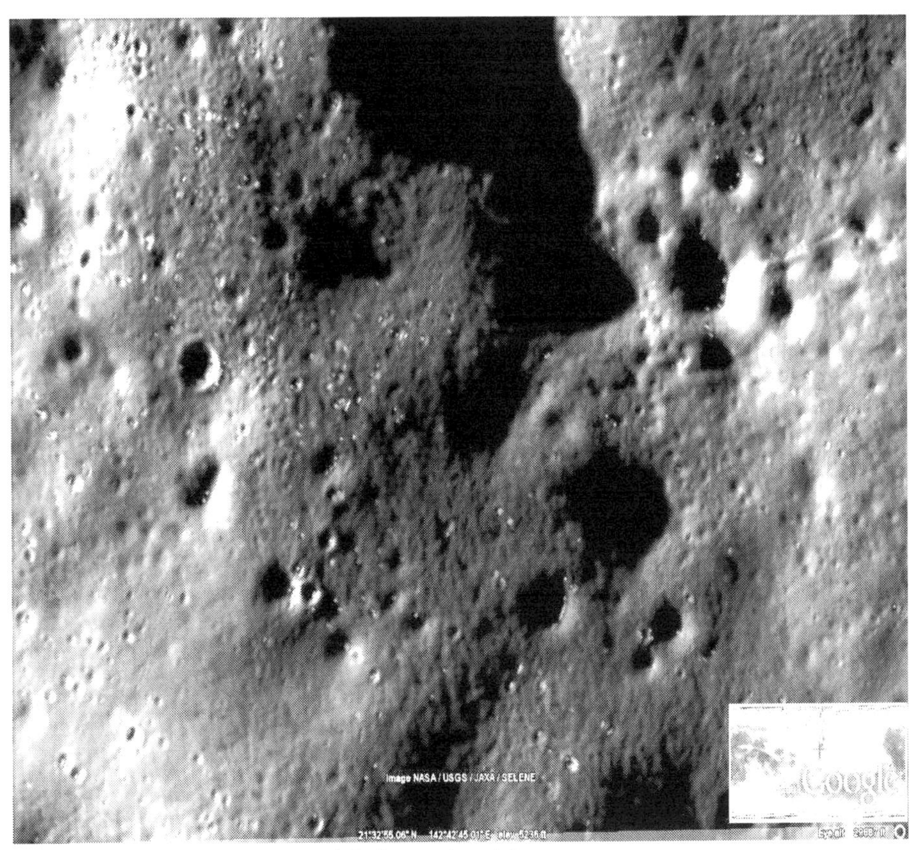

High contrast image

Image NASA / USGS / JAXA / SELENE
28°42'07.41" N 147°44'43.49" E
Eye alt 4955 ft

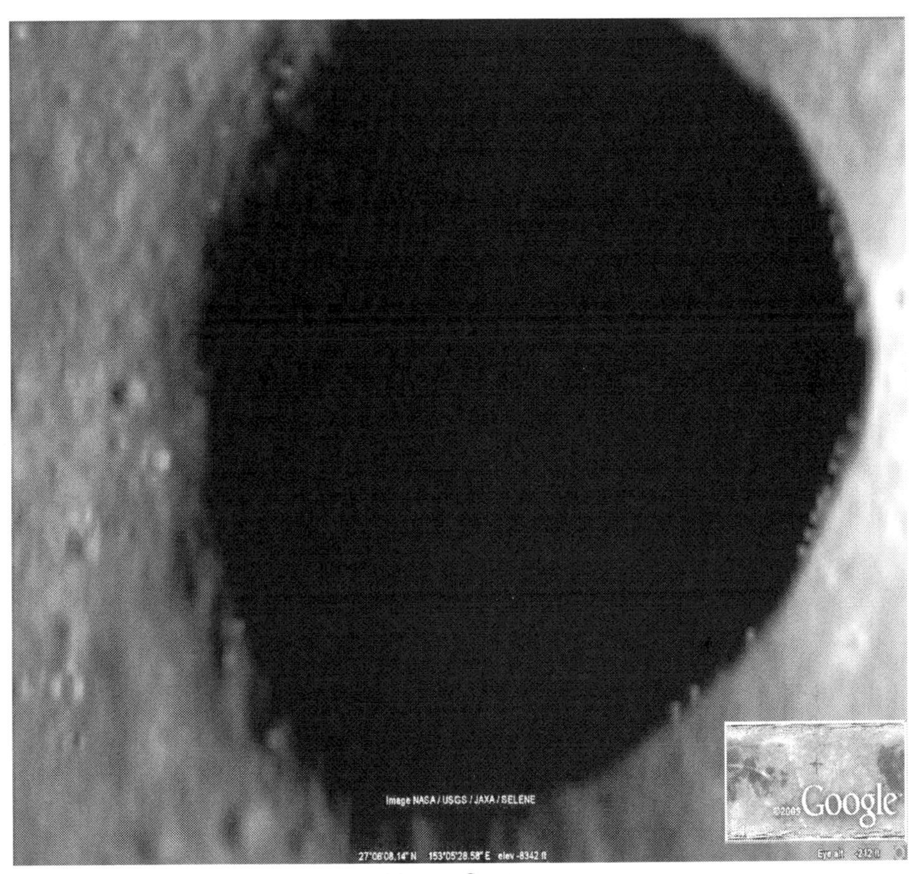
Image NASA / USGS / JAXA / SELENE
Google
27°08'08.14" N 153°05'28.58" E elev -8342 ft

Huge Crater

Image NASA / USGS / JAXA / SELENE
©2009 Google
29°28'02.91" N 143°19'49.55" E
Eye alt 6022 ft

Closer view of a moon city

Dec 31, 2003 1:55 am
Dec 30 2003
Dec 31, 2003
Untitled Placemark
Google

UFO Valley

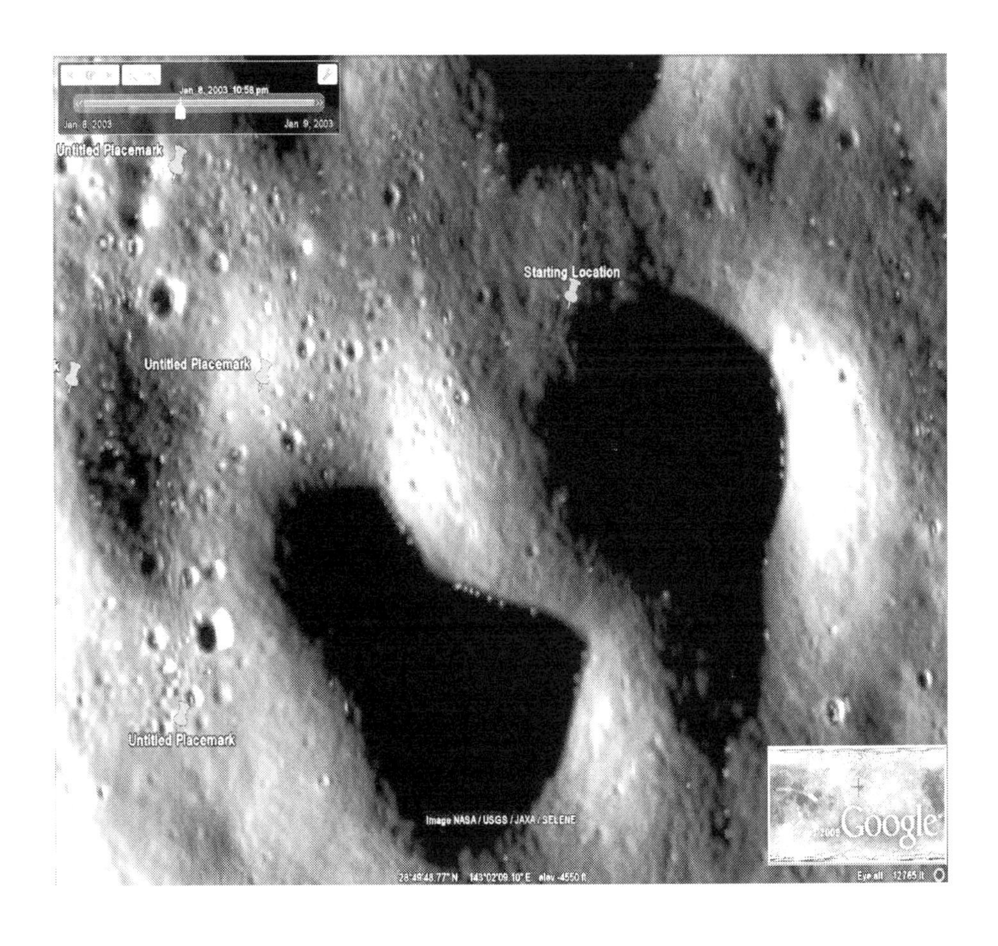
Jan 8, 2003 10:58 pm
Jan 8, 2003
Jan 9, 2003
Untitled Placemark
Starting Location
Untitled Placemark
Untitled Placemark
Image NASA / USGS / JAXA / SELENE
Google
Eye alt 12765 ft

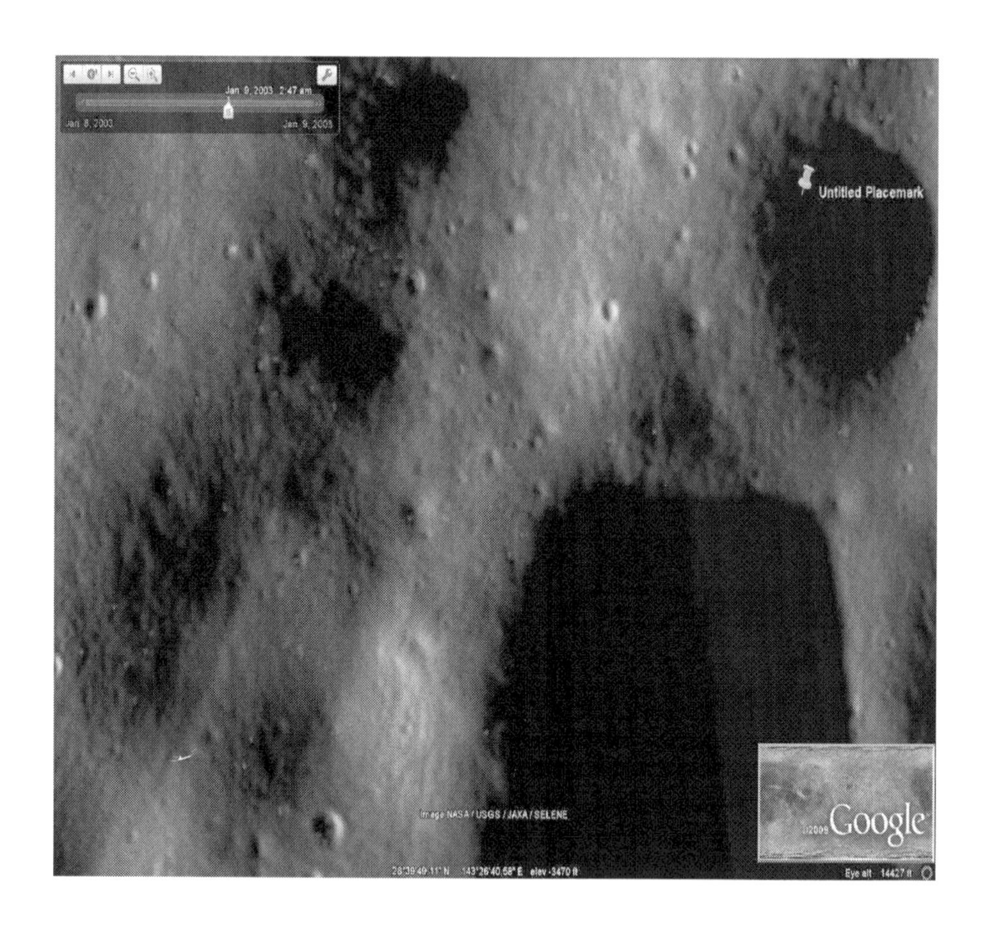
Jan 9, 2003 2:47 am
Jan 8, 2003
Jan 9, 2003
Untitled Placemark
Image NASA / USGS / JAXA / SELENE
©2009 Google
28°39'49.11" N 143°26'40.58" E elev -3470 ft
Eye alt 14427 ft

Image NASA / USGS / JAXA / SELENE
Google earth

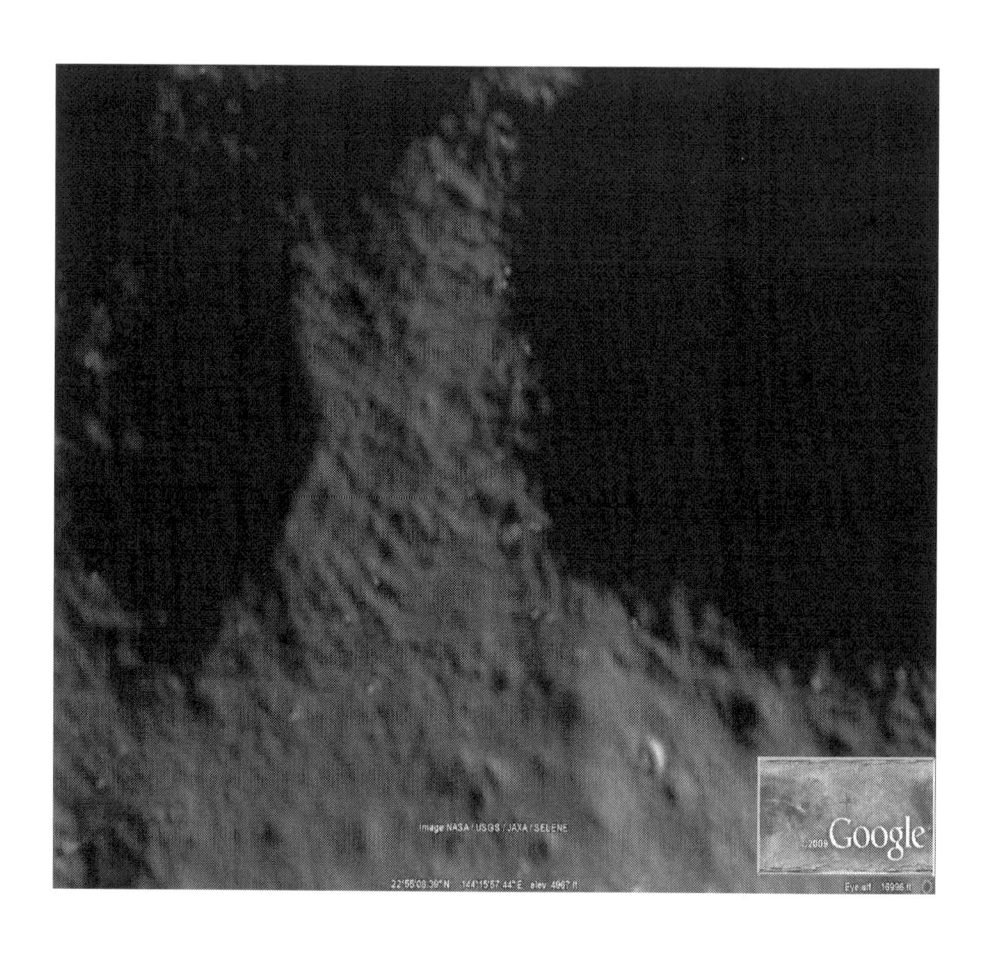
Image NASA / USGS / JAXA / SELENE
Google

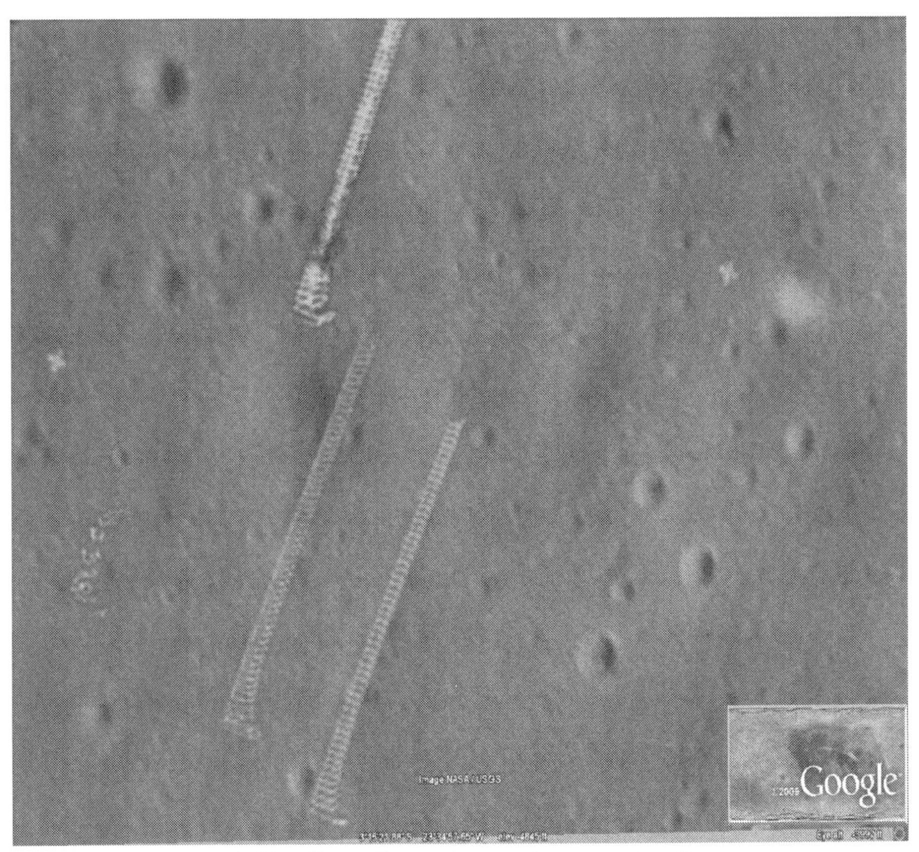

3 ¼ mile coils and one damaged coil

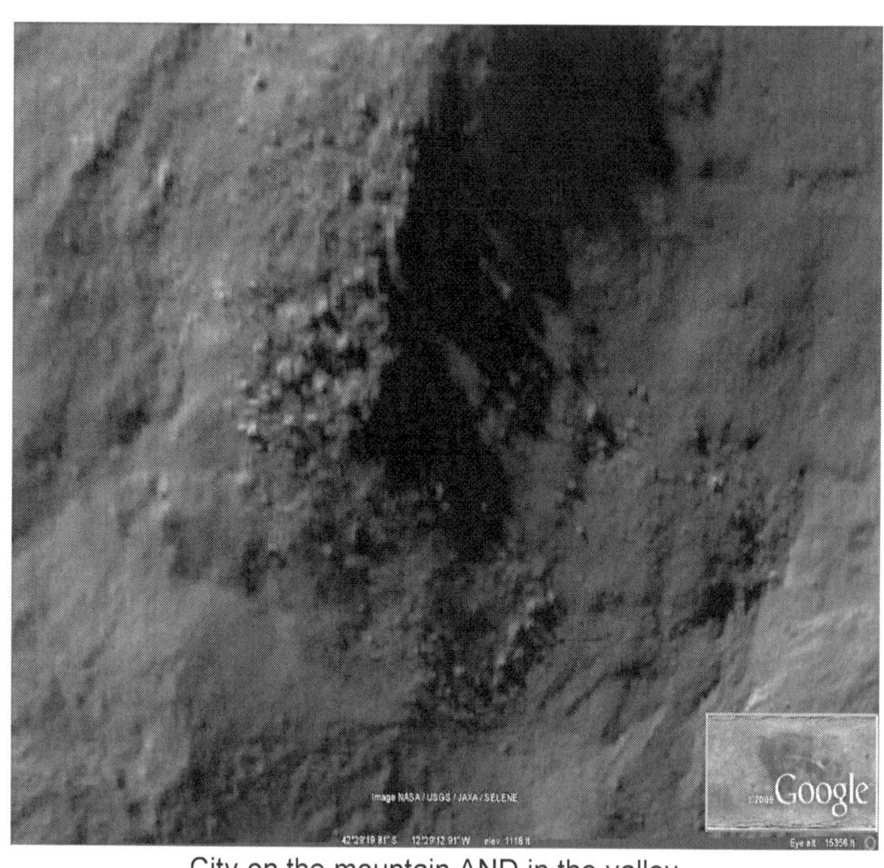

City on the mountain AND in the valley

Moons surface without enhancement

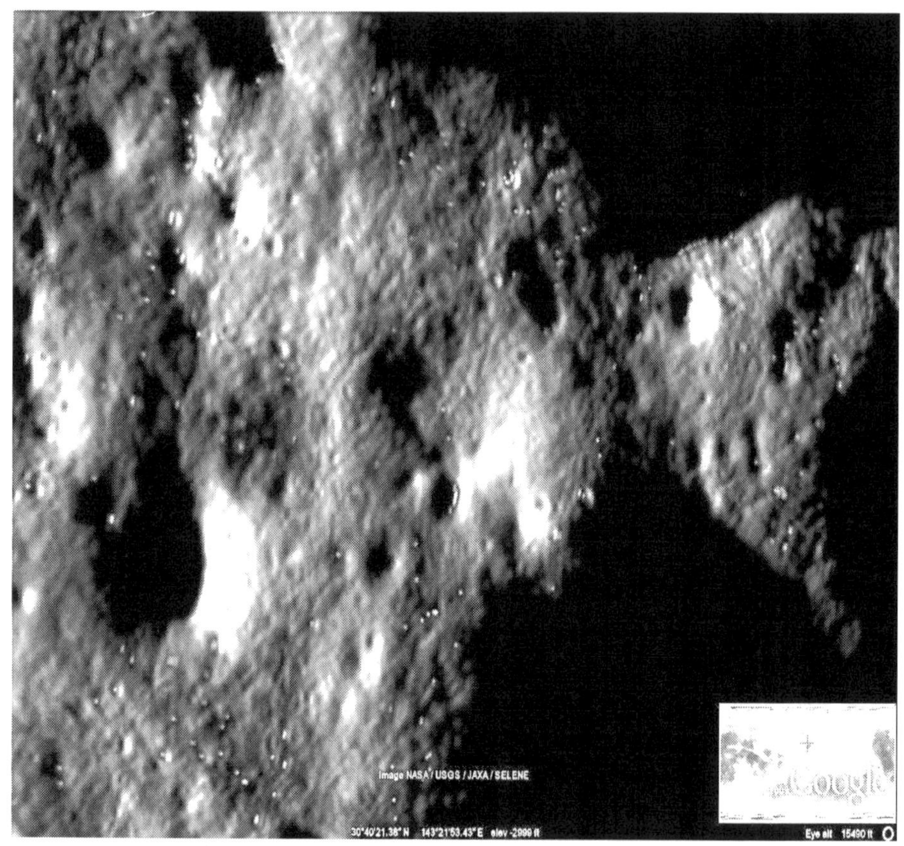

Moons surface with enhancement

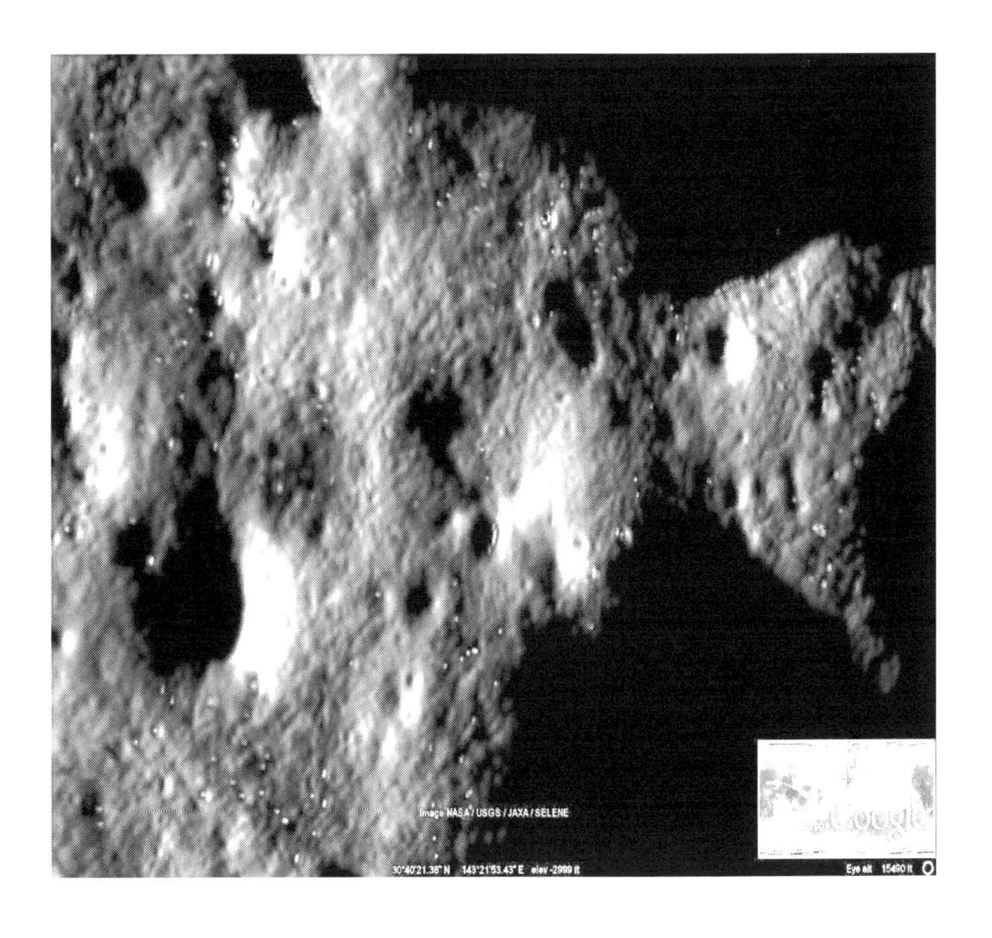
Image NASA / USGS / JAXA / SELENE
30°40'21.38" N 143°21'53.43" E elev -2999 ft
Eye alt 15490 ft

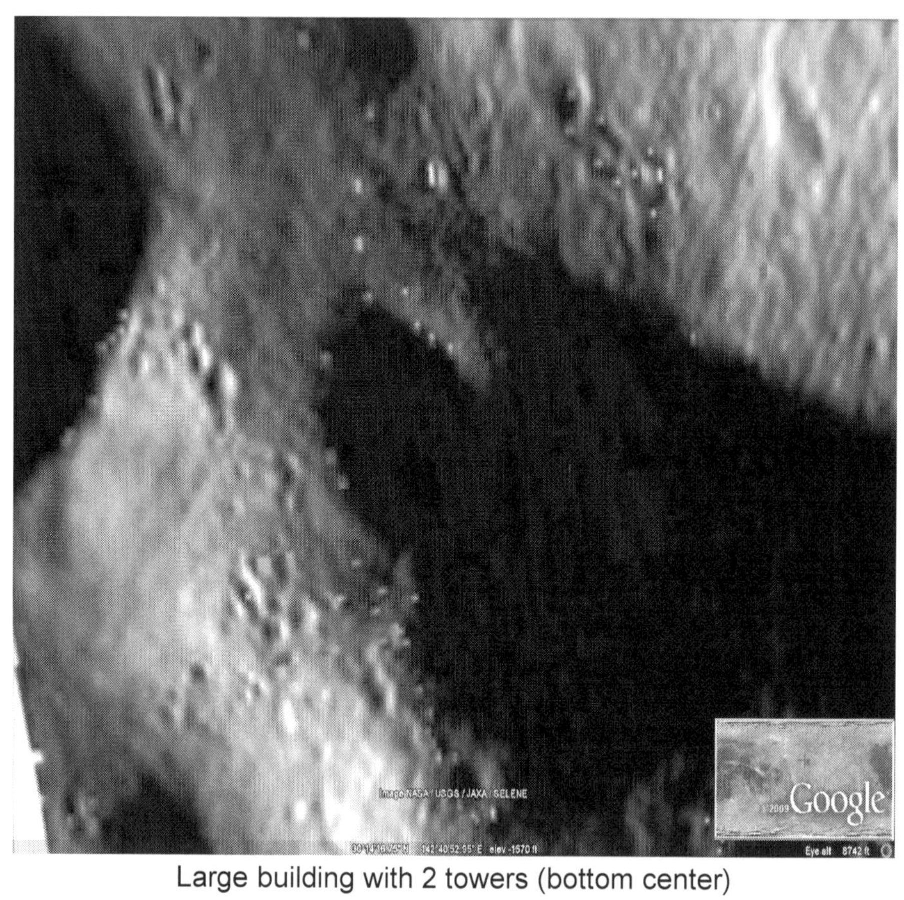

Large building with 2 towers (bottom center)

Dec 31, 2003 3:55 am
Untitled Placemark
Image NASA / USGS / JAXA / SELENE
Google
29°32'35.42" N 144°21'28.97" E elev -5031 ft
Eye alt 12939 ft

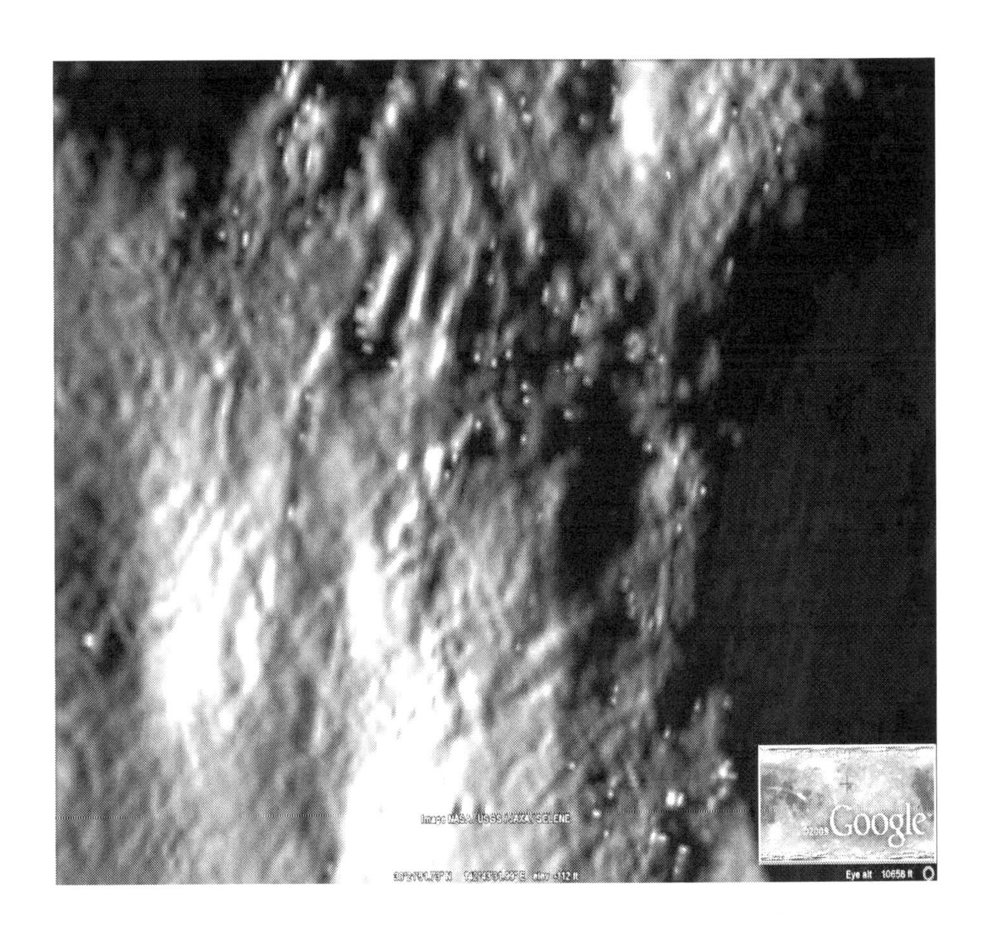

Buildings in bottom right corner

Image NASA / USGS / JAXA / SELENE
21°31'56.26" N 142°42'15.55" E elev 5519 ft
Eye alt 21340 ft

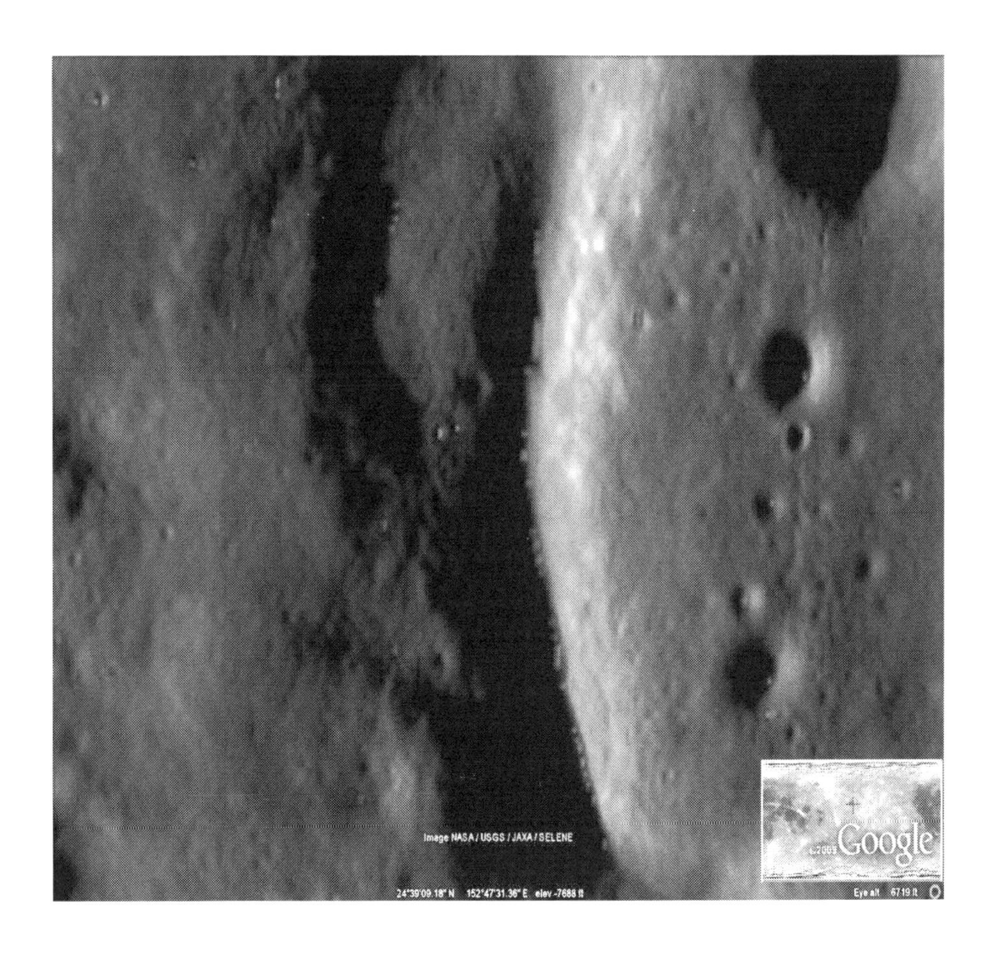
Image NASA / USGS / JAXA / SELENE
Google
24°39'09.18" N 152°47'31.36" E elev -7688 ft
Eye alt 6719 ft

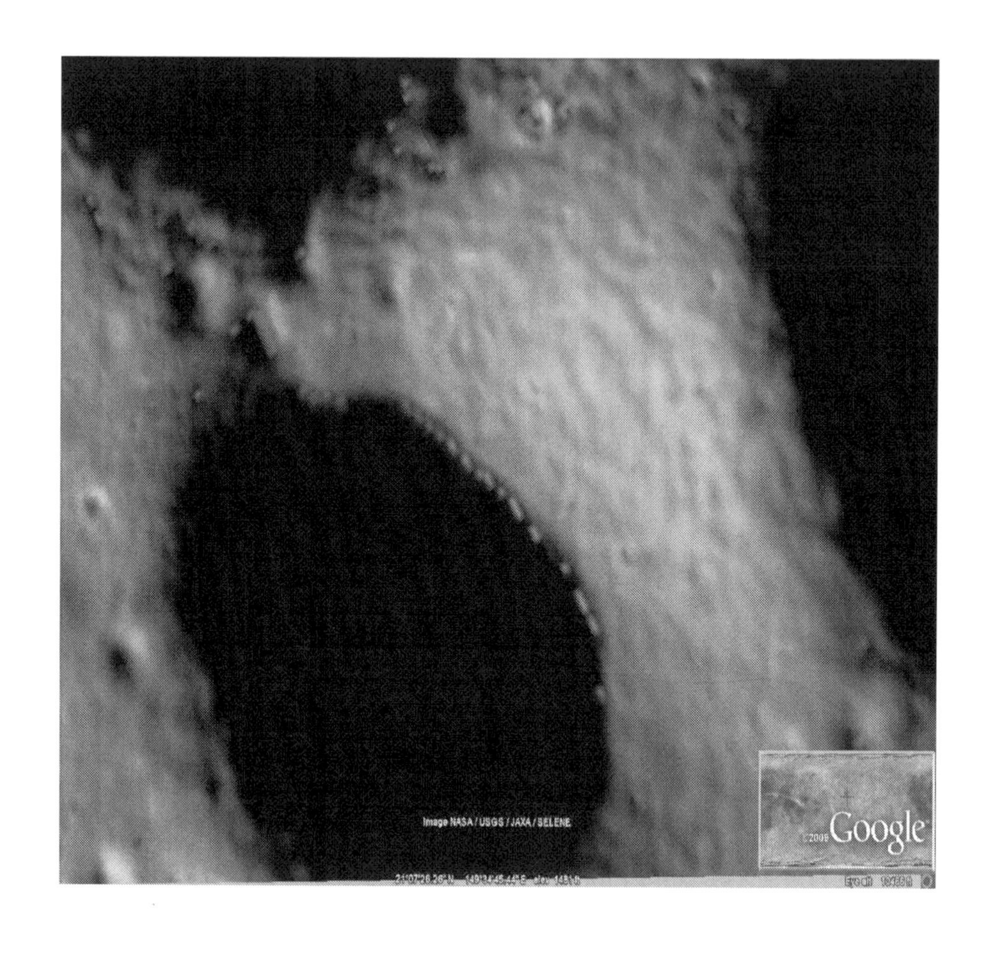
Image NASA / USGS / JAXA / SELENE
©2009 Google

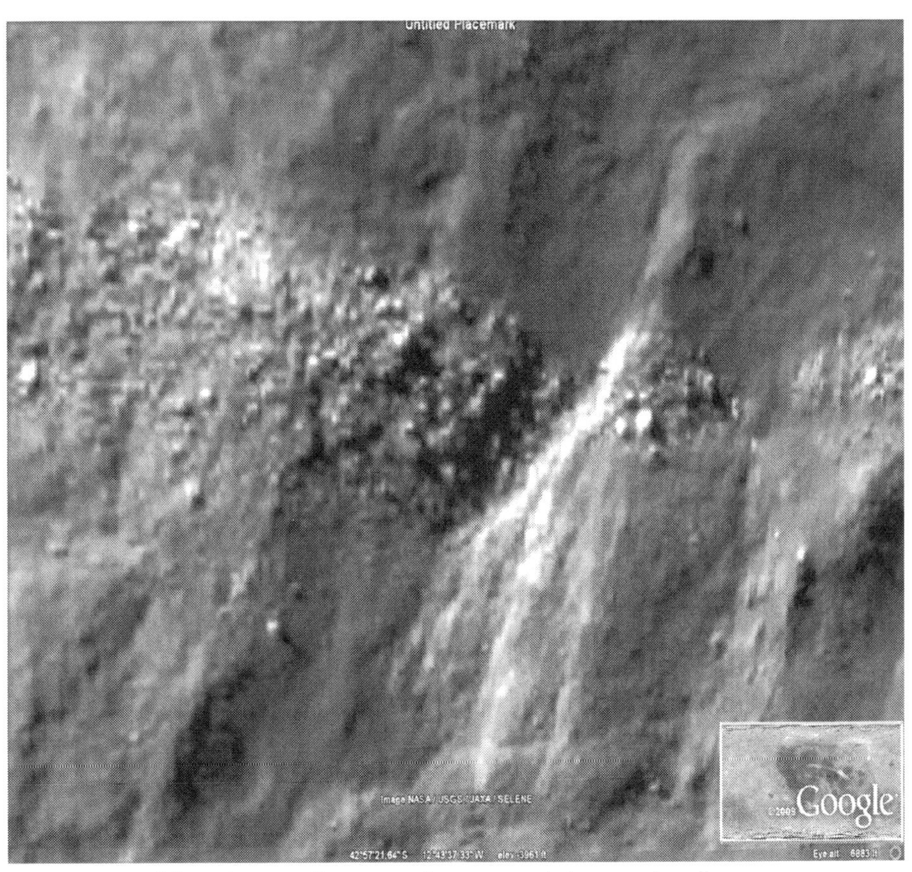

Very large city spanning mountains and valleys

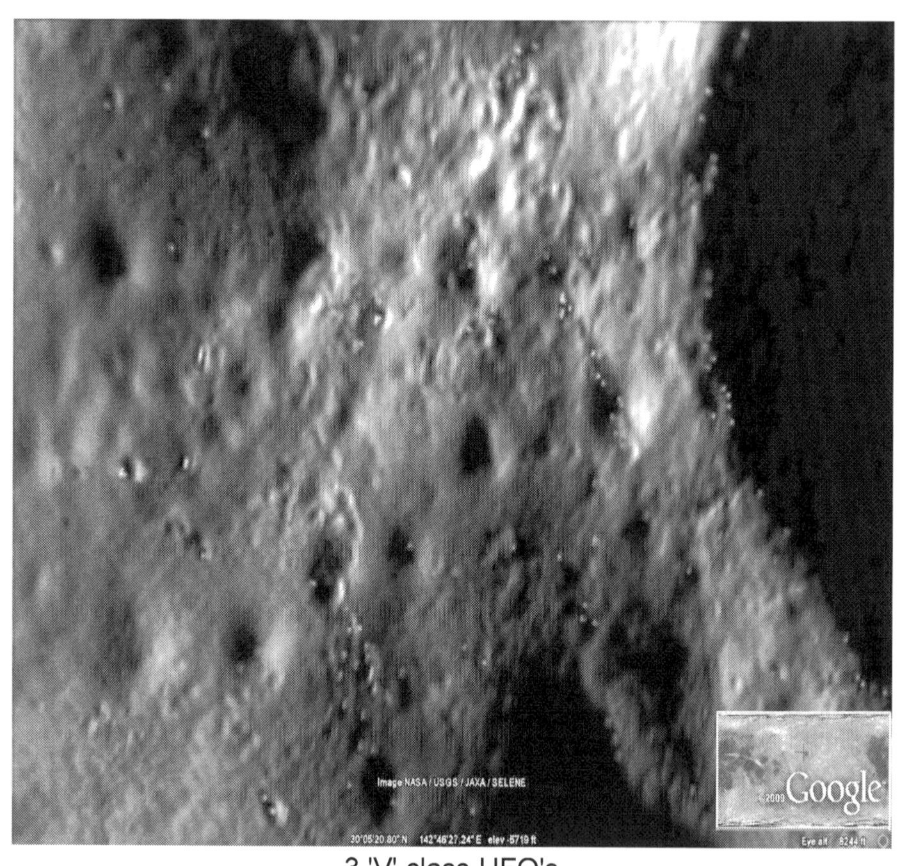

3 'V' class UFO's

EGG UFO BIGEST
Image NASA / USGS / JAXA / SELENE
Google earth

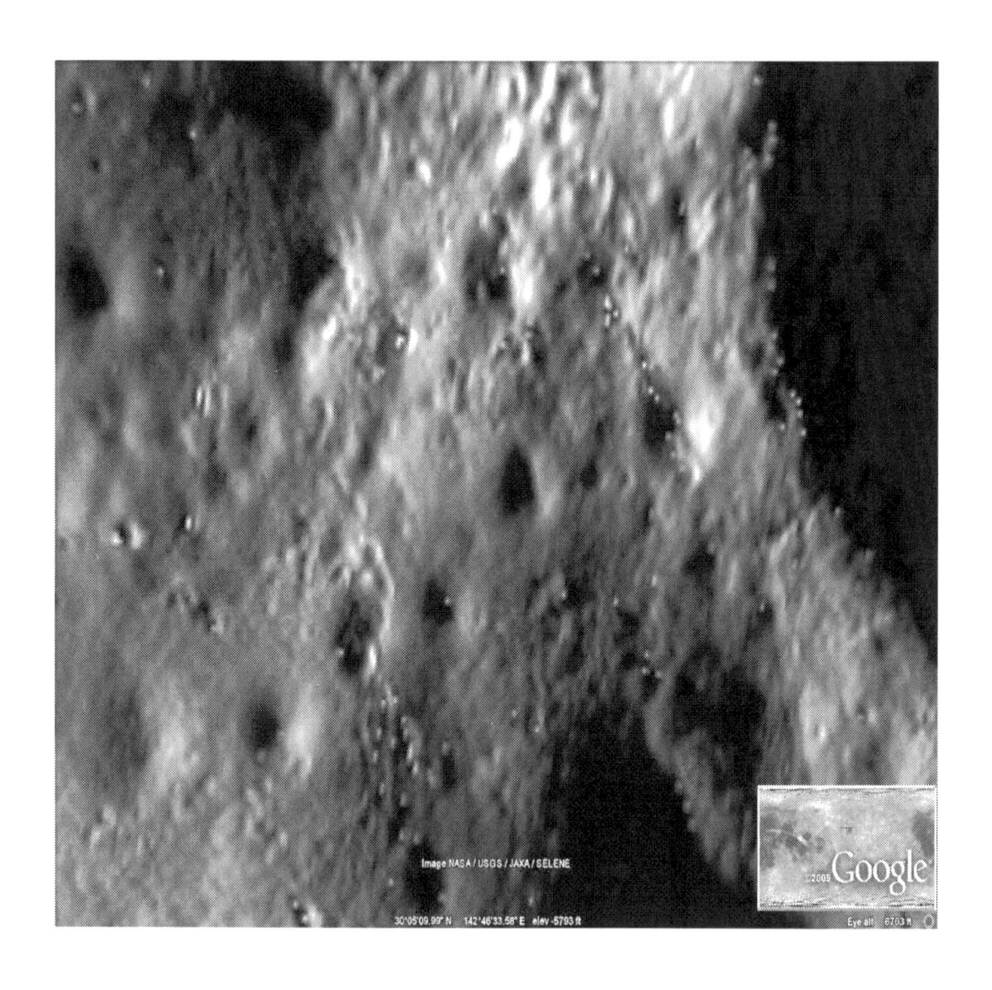
Image NASA / USGS / JAXA / SELENE
©2009 Google
30°05'09.99" N 142°46'33.58" E elev -5793 ft
Eye alt 6703 ft

'V' Class ship in center of page on small crater

Image NASA / USGS / JAXA / SELENE
Google

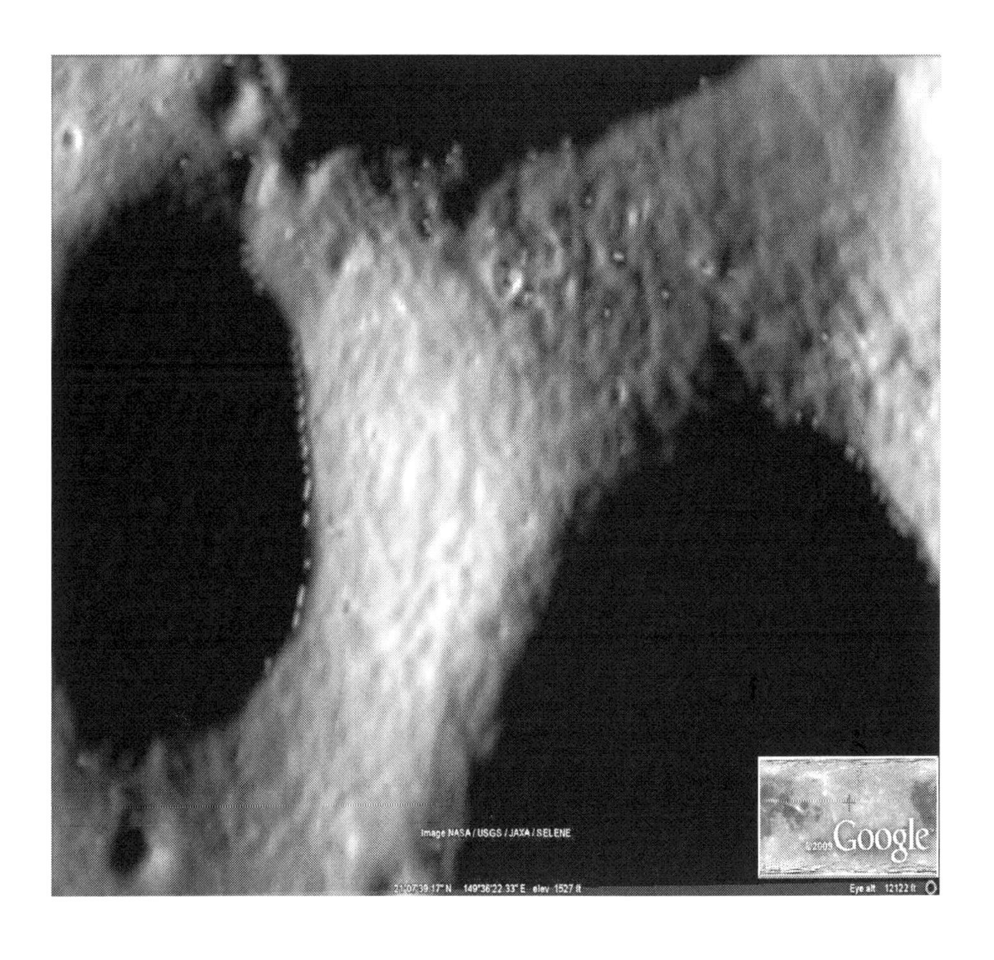
Image NASA / USGS / JAXA / SELENE
©2009 Google
21°07'39.17" N 149°36'22.33" E elev 1527 ft
Eye alt 12122 ft

Image NASA / USGS / JAXA / SELENE
Google
142°58'41.52" E elev -4462 ft
Eye alt 6082 ft

I

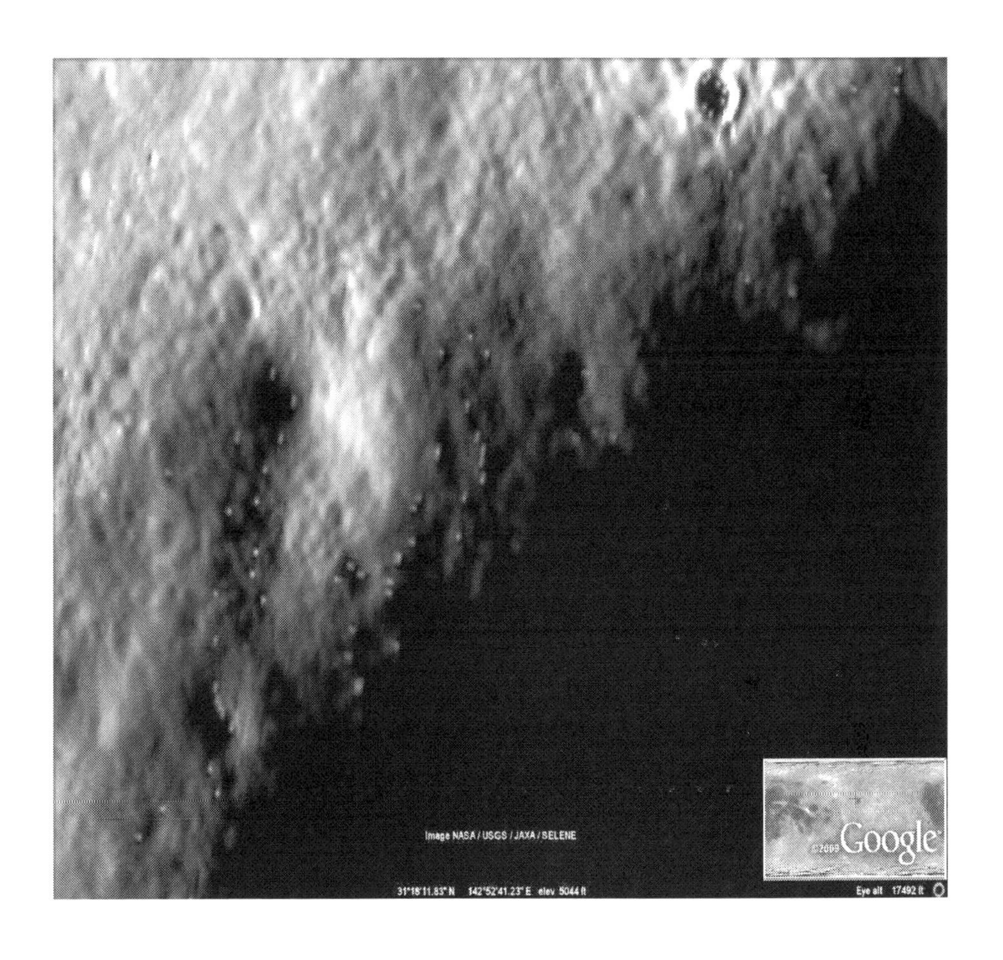
Image NASA / USGS / JAXA / SELENE
©2009 Google
31°18'11.83" N 142°52'41.23" E elev 5044 ft
Eye alt 17492 ft

Image NASA / USGS / JAXA / SELENE
Google
30°10'35.84" N 142°48'21.55" E

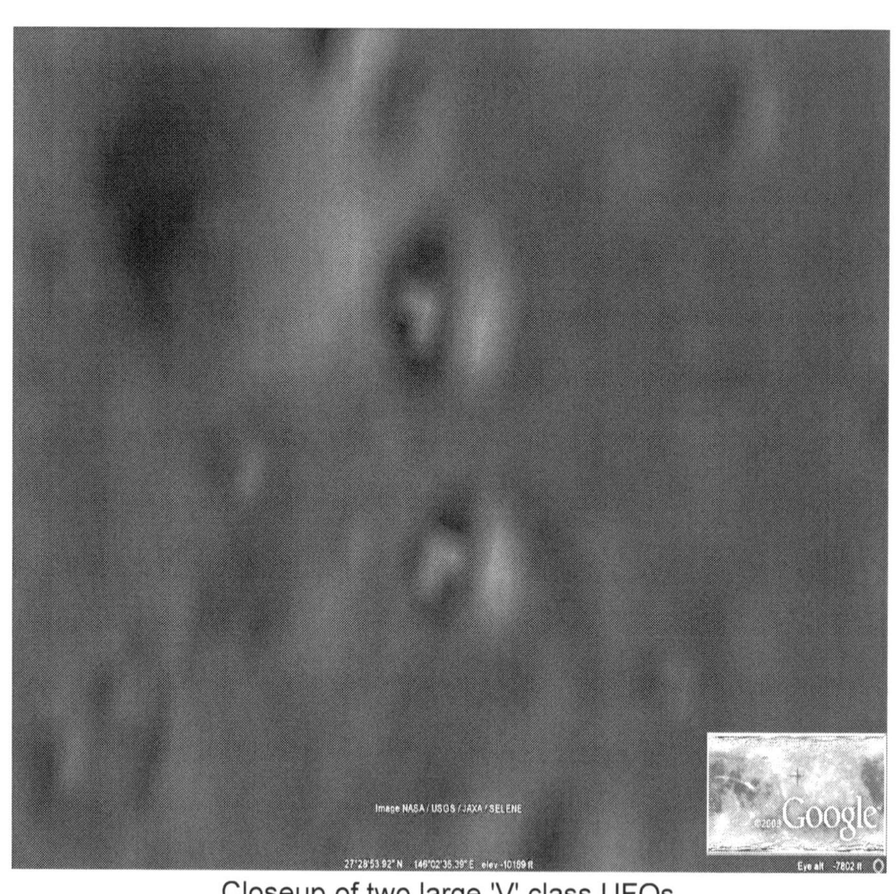

Closeup of two large 'V' class UFOs

Same ships from distance

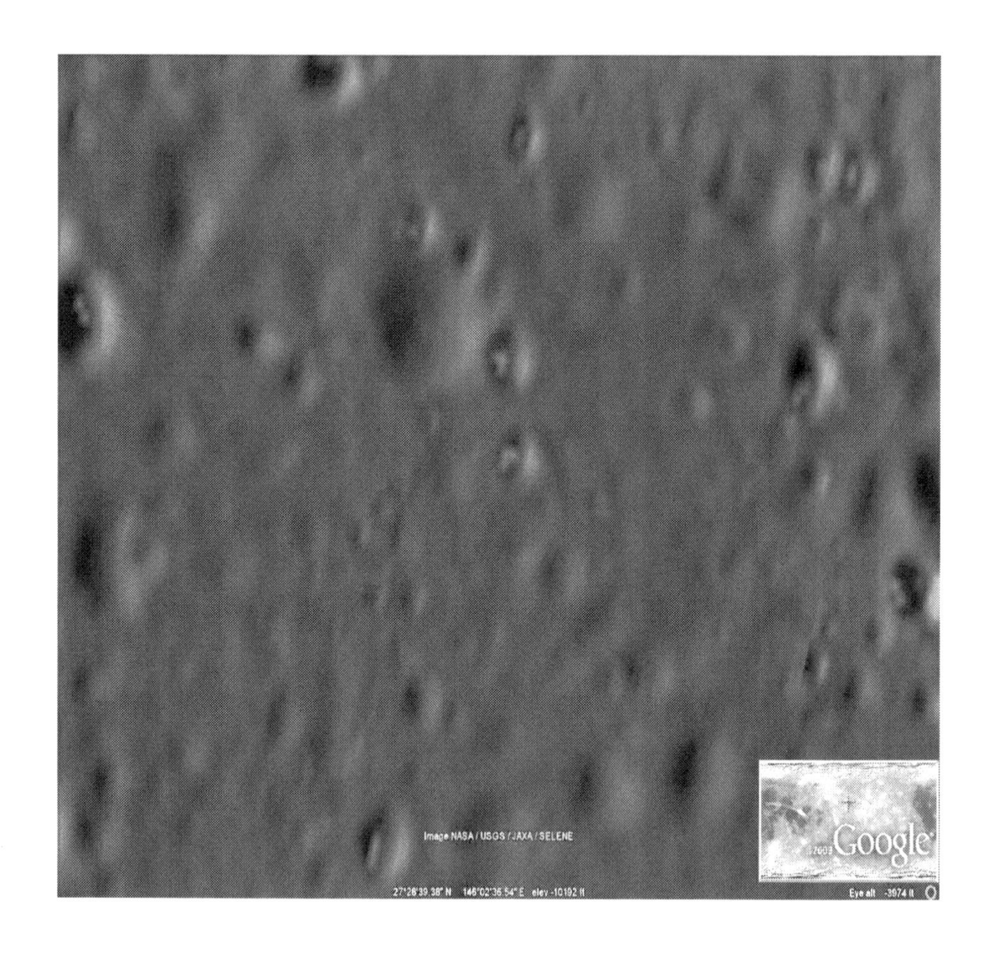
Image NASA / USGS / JAXA / SELENE
Google
27°28'39.38" N 146°02'36.54" E elev -10192 ft
Eye alt -3974 ft

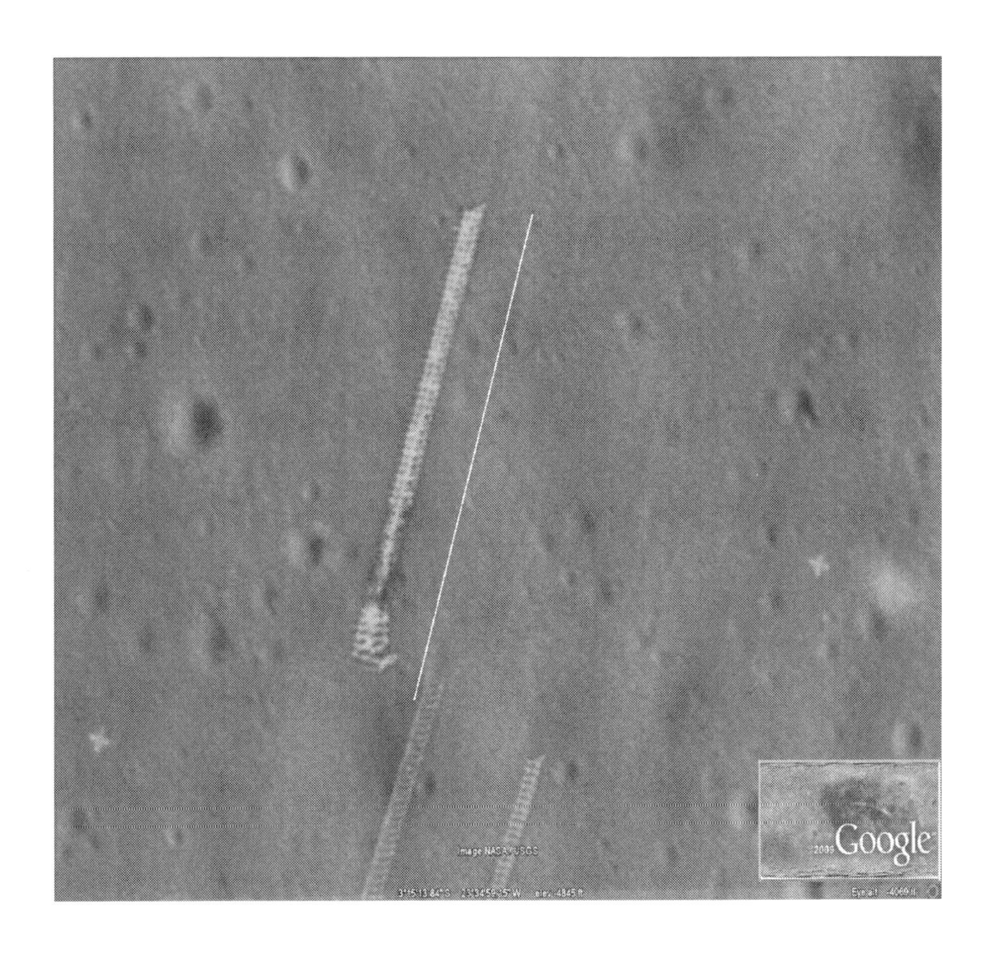
Image NASA/USGS
Google

Image NASA / USGS / JAXA / SELENE
28°43'18.55" N 143°24'16.42" E
Google
Eye alt 9519 ft

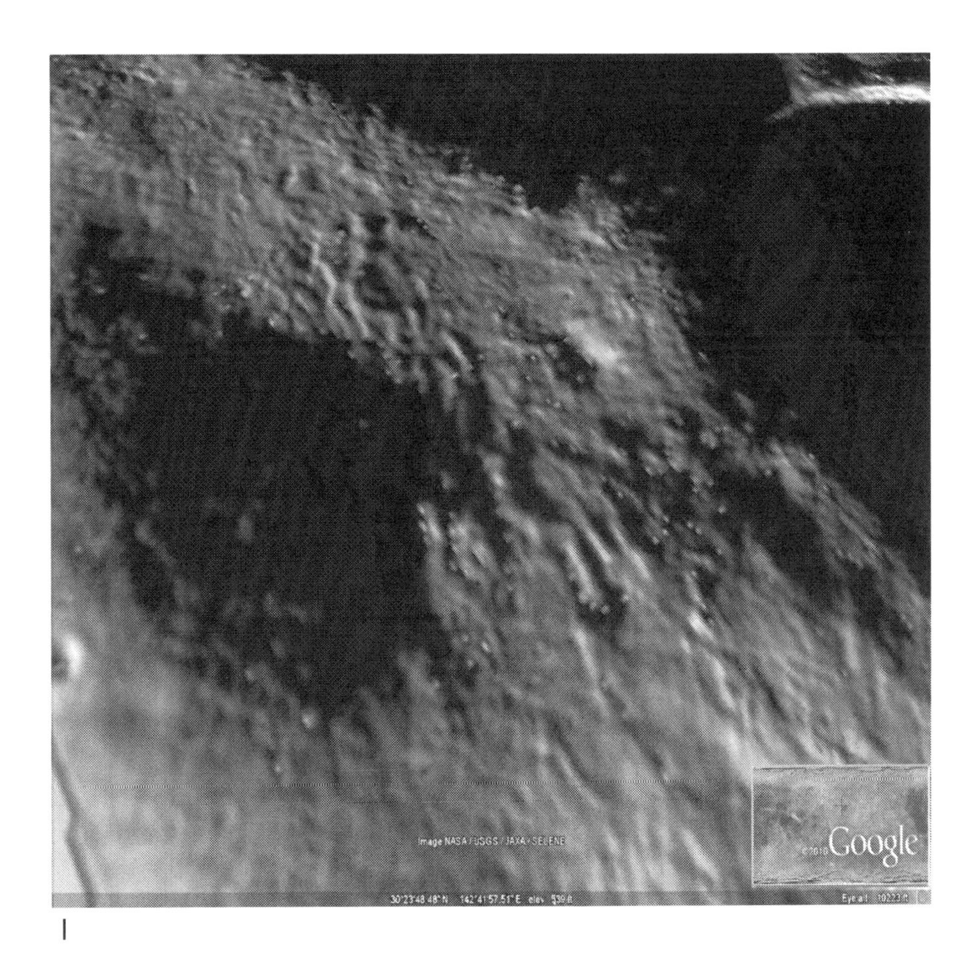
Image NASA/USGS/JAXA/SELENE
Google
30°23'48.48" N 142°41'57.51" E elev 539 ft
Eye alt 10223 ft

Image NASA / USGS / JAXA / SELENE
Google
29°44'19.92" N 143°10'55.88" E elev -6234 ft
Eye alt 6975 ft

Image NASA / USGS / JAXA / SELENE
Google
30°06'50.90" N 143°15'11.97" E elev -5439 ft
Eye alt 17263 ft

Image NASA / USGS / JAXA / SELENE
©2010 Google
30°28'36.62" N 143°06'00.15" E elev -2290 ft
Eye alt 16545 ft

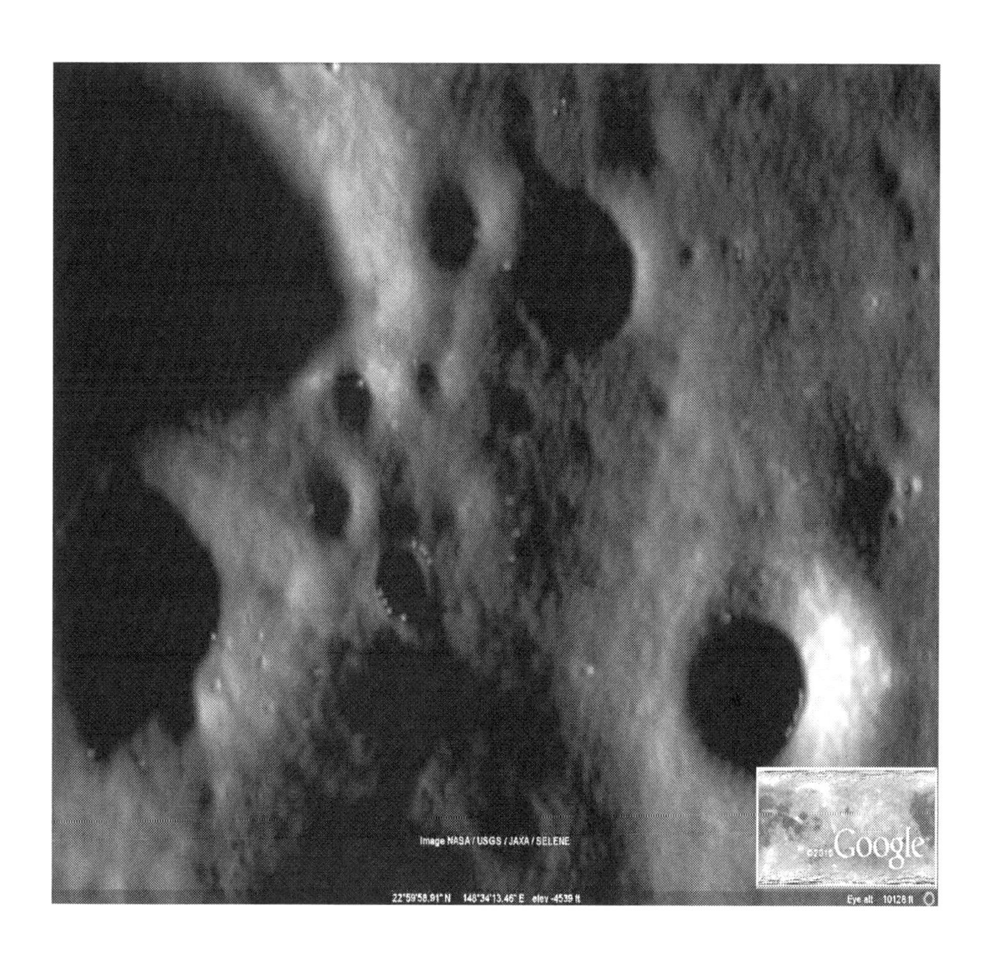
Image NASA / USGS / JAXA / SELENE
Google
22°59'58.91" N 148°34'13.46" E elev -4539 ft
Eye alt 10128 ft

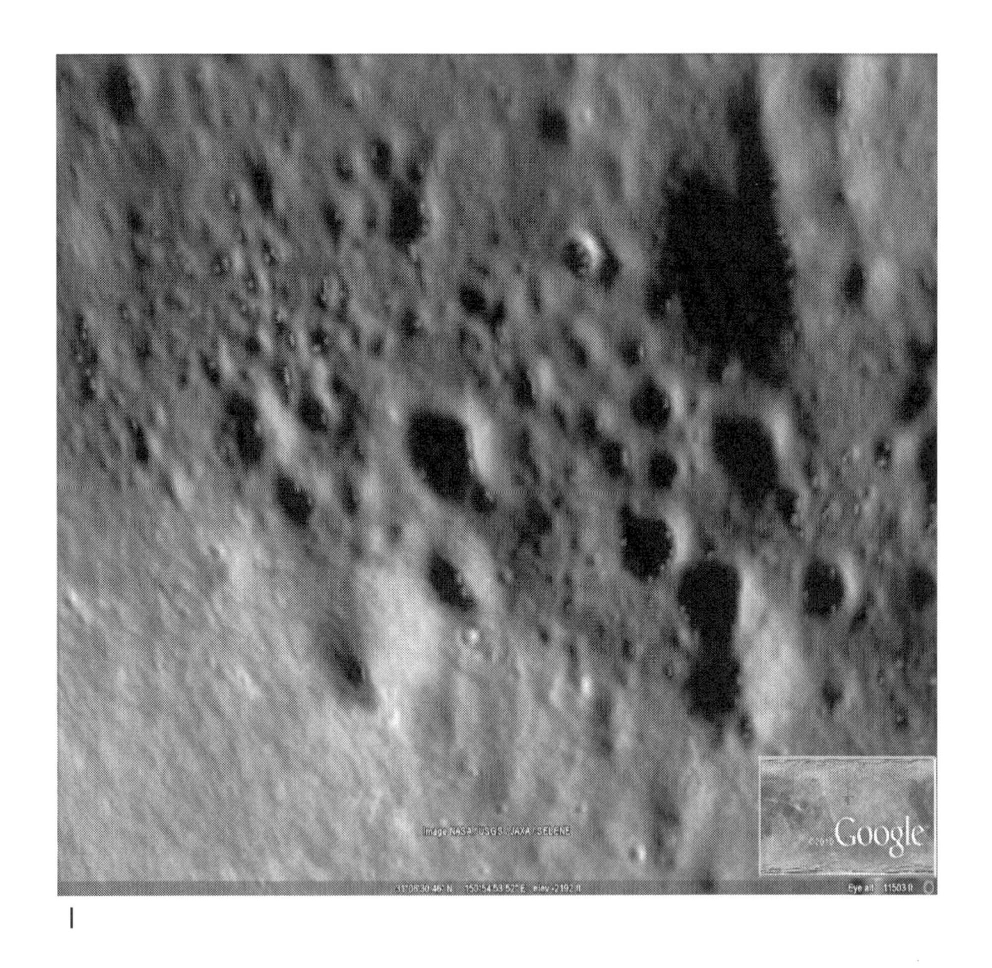
Image NASA / USGS / JAXA / SELENE
Google
Eye alt 11503 ft

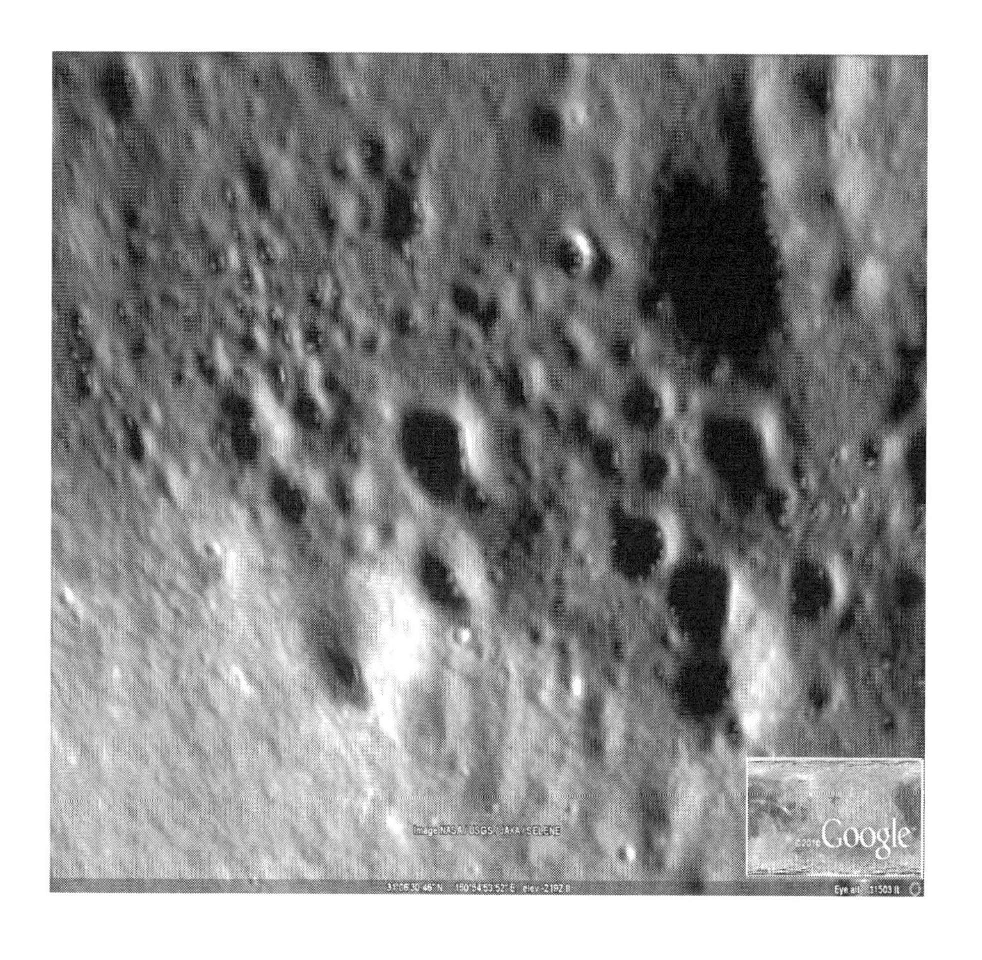
Image NASA / USGS / JAXA / SELENE
Google

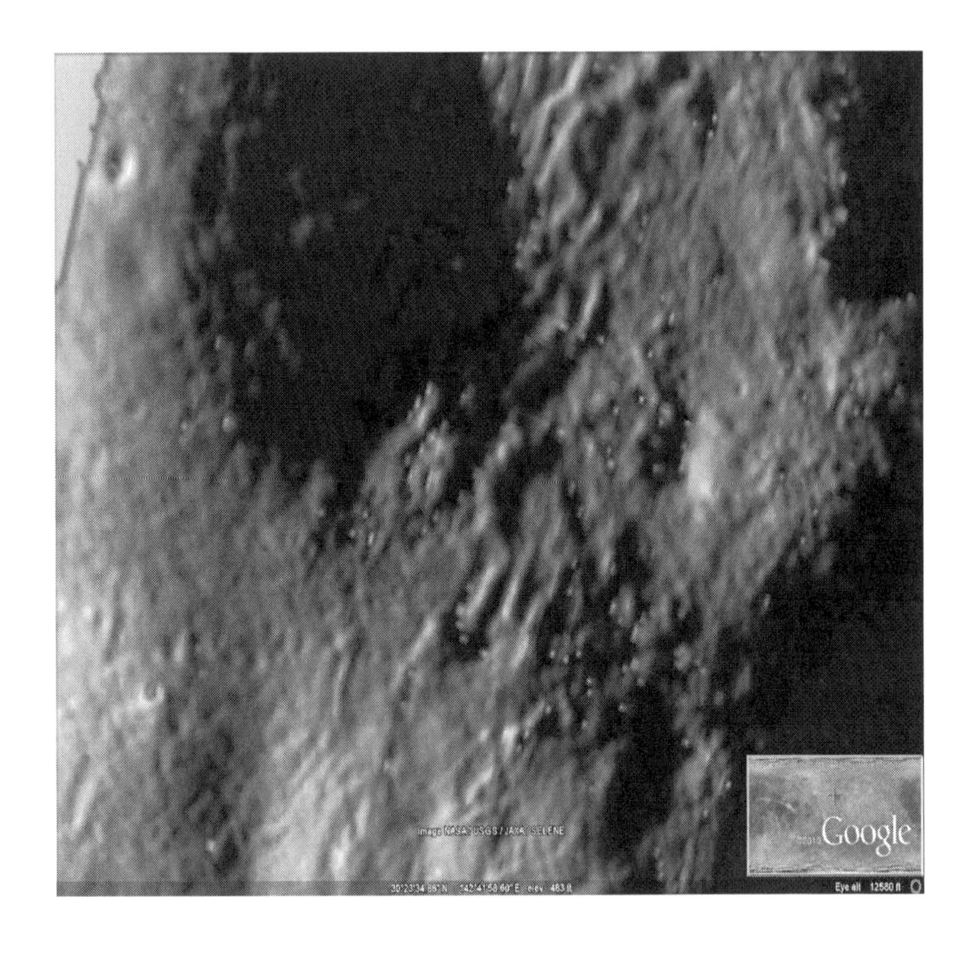
Image NASA / USGS / JAXA / SELENE
Google
elev 483 ft
Eye alt 12580 ft

11 cigar shaped UFOs together

Image NASA / USGS / JAXA / SELENE
29°53'56.42" N 145°21'57.51" E elev -5343 ft
Eye alt 11024 ft
Google

Image NASA / USGS / JAXA / SELENE
Google
Eye alt 3939 ft

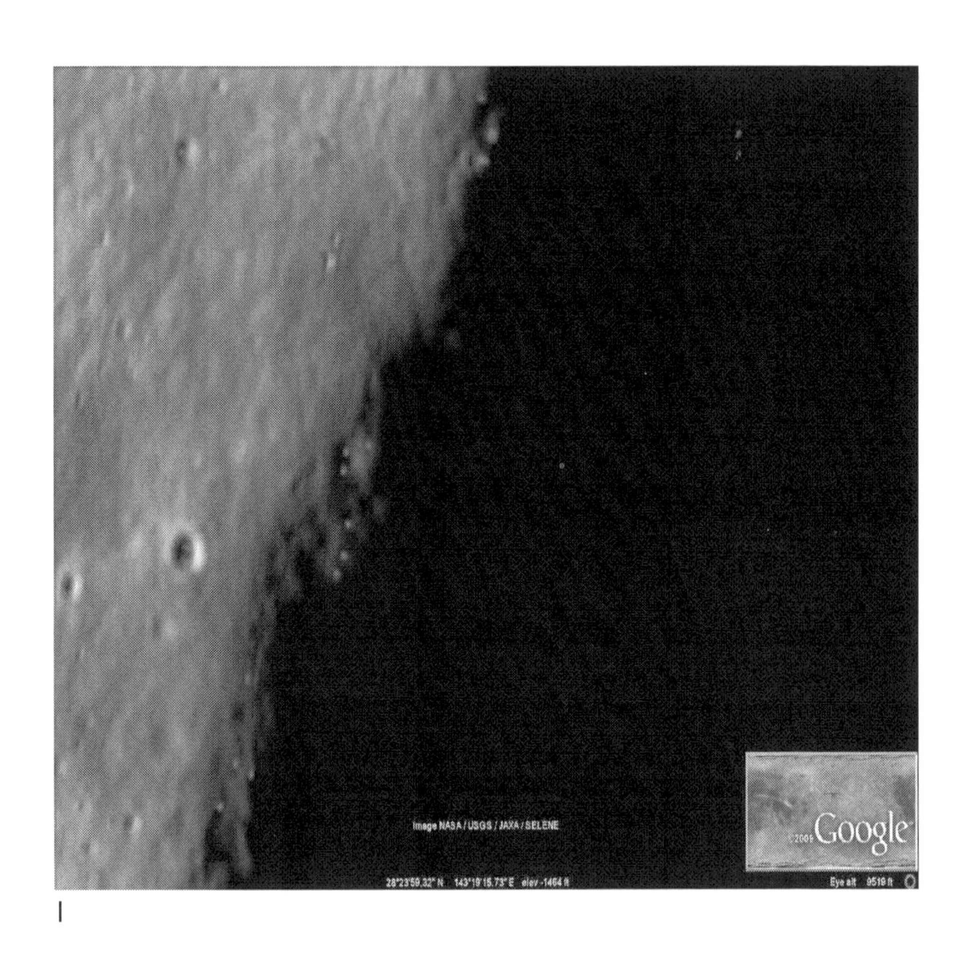
Image NASA / USGS / JAXA / SELENE
Google
28°23'59.32" N 143°19'15.73" E elev -1464 ft
Eye alt 9519 ft

I

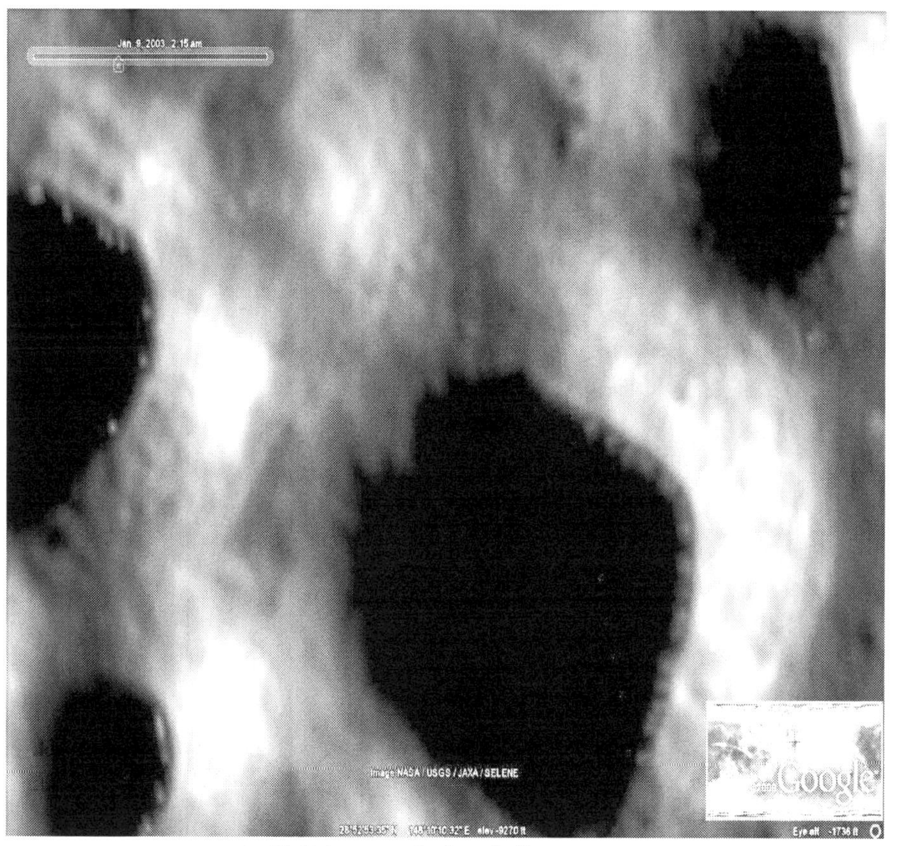

Tall towers in top left corner

Image NASA / USGS / JAXA / SELENE
©2009 Google
29°28'02.91" N 143°19'49.55" E elev -6217 ft
Eye alt 6022 ft

Image NASA / USGS / JAXA / SELENE
Google
Eye alt 14520 ft

Structures

Huge building center right, light on center left

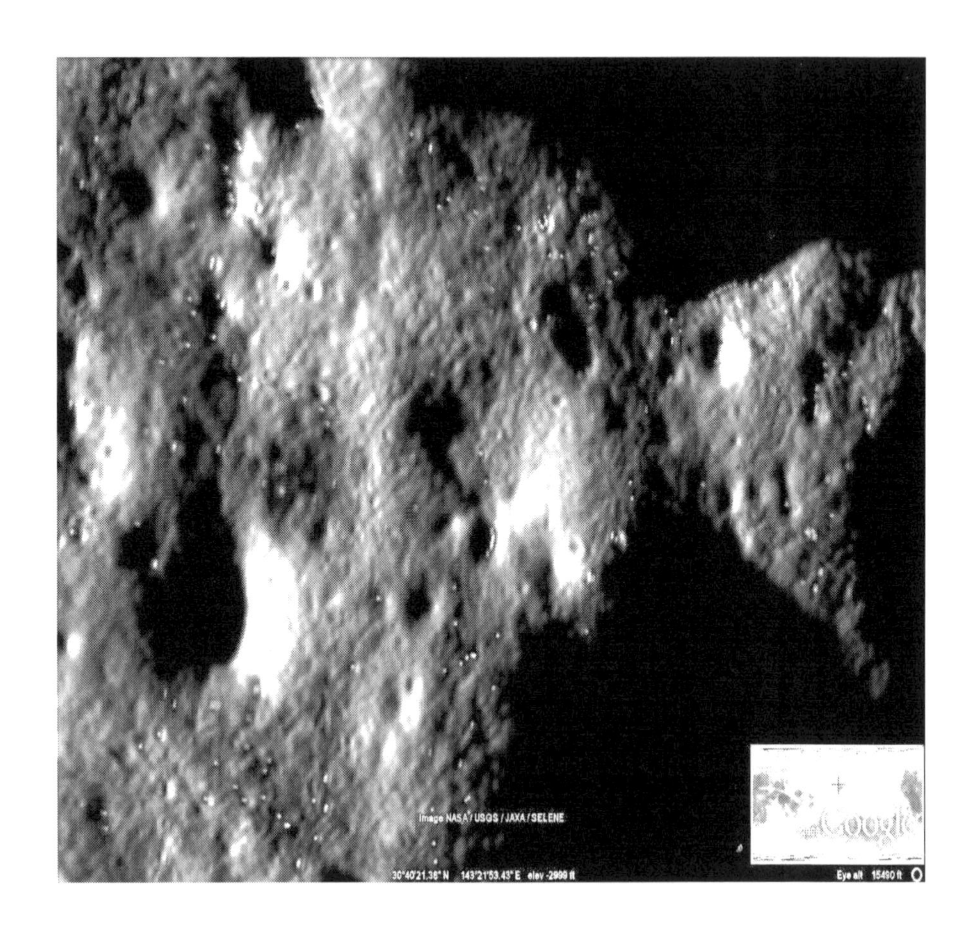
Image NASA / USGS / JAXA / SELENE
Google
30°40'21.36" N 143°21'53.43" E elev -2999 ft
Eye alt 15490 ft

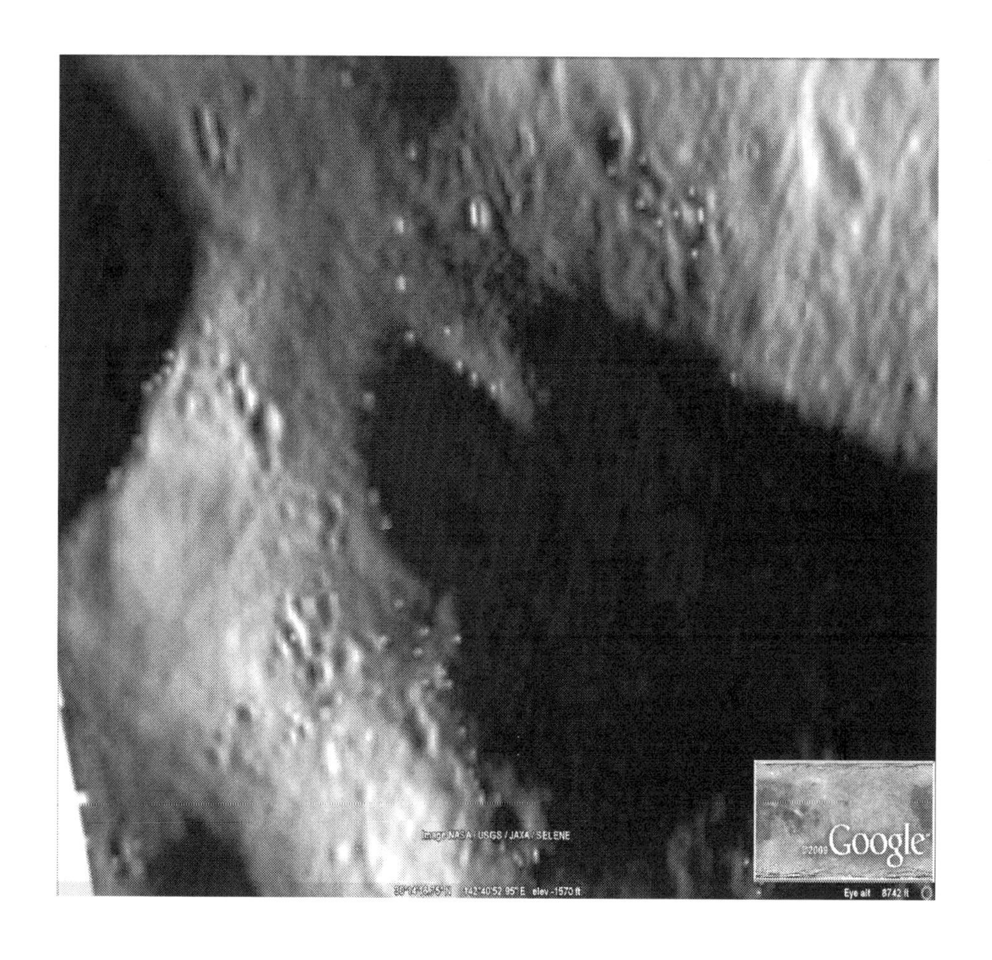
Image NASA / USGS / JAXA / SELENE
©2009 Google
142°40'52.95" E elev -1570 ft
Eye alt 8742 ft

Dec 31, 2003 3:31 am
Image NASA / USGS / JAXA / SELENE
Google

Dec 31, 2003 3:55 am
Untitled Placemark
Image NASA / USGS / JAXA / SELENE
Google
29°32'35.42" N 144°21'28.97" E elev -5031 ft
Eye alt 12939 ft

Image NASA / USGS / JAXA / SELENE
©2009 Google
29°52'19.11" N 143°51'32.30" E elev 5957 ft
Eye alt 9585 ft

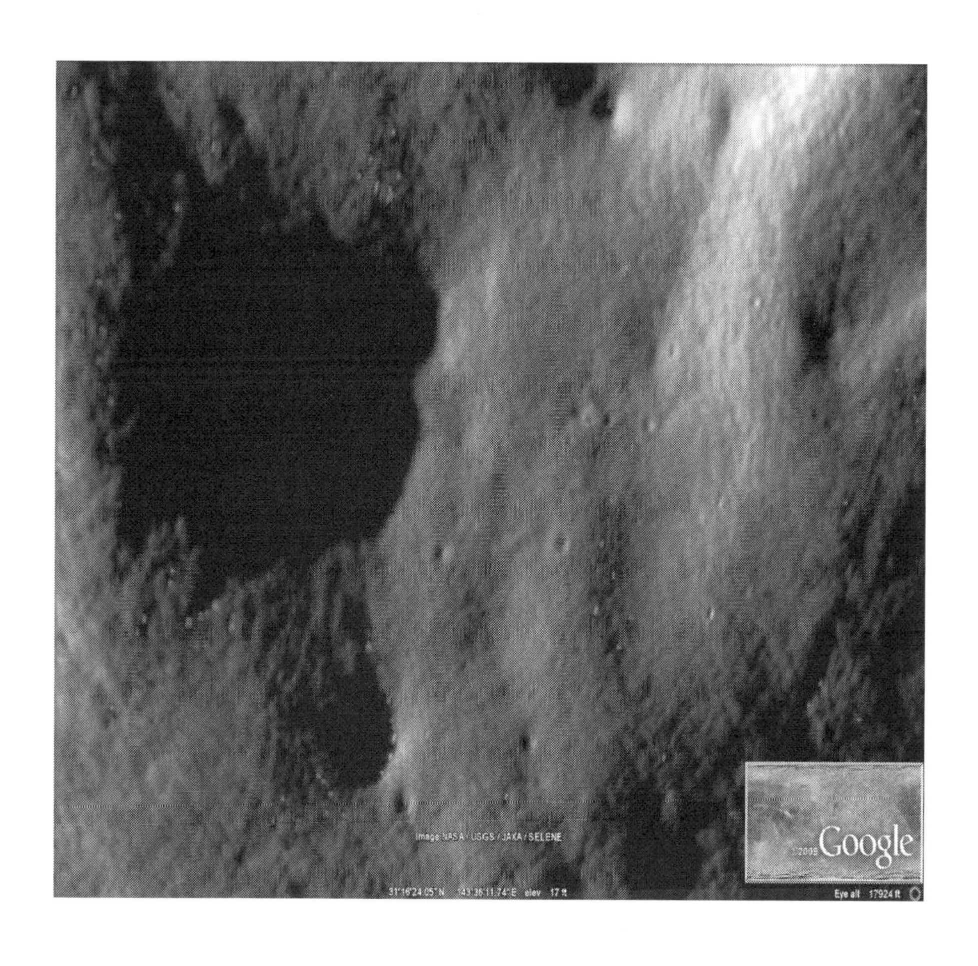
Image NASA / USGS / JAXA / SELENE
Google
31°16'24.05" N 143°36'11.74" E elev 17 ft
Eye alt 17924 ft

Image NASA / USGS / JAXA / SELENE
22°42'26.45" N 142°48'59.30" E elev -3359 ft

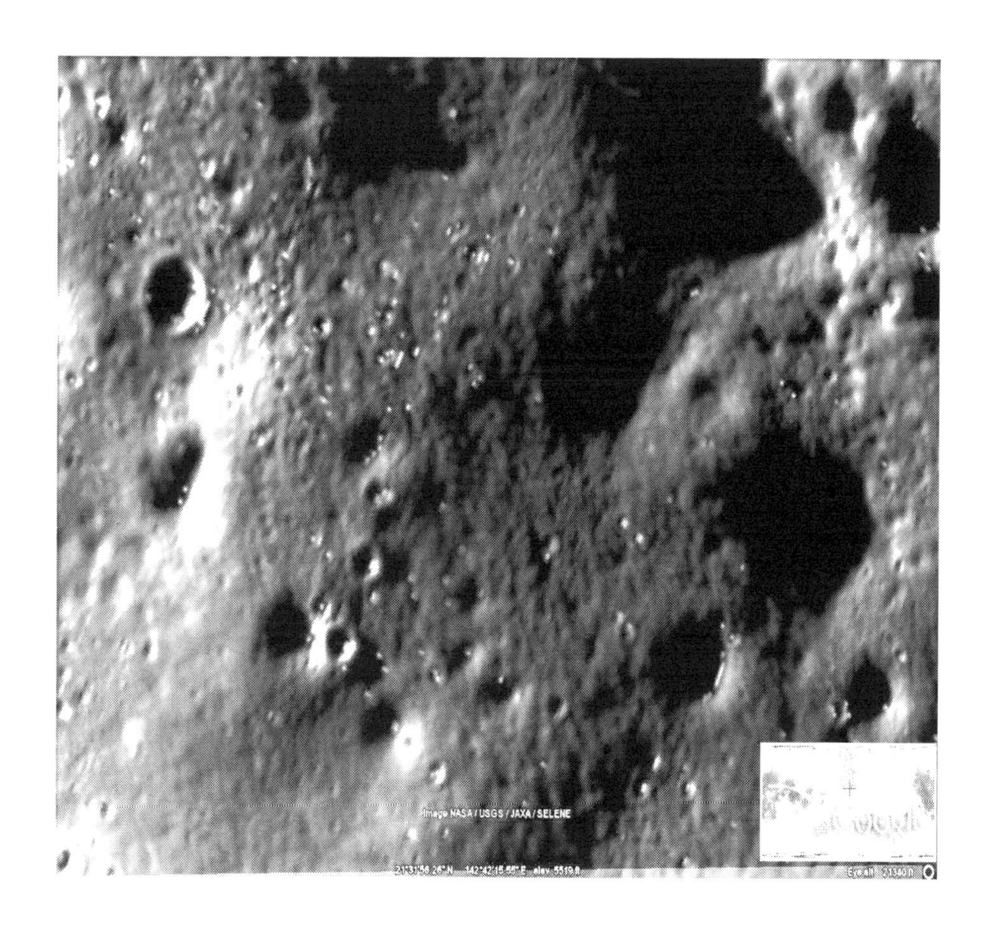
Image NASA / USGS / JAXA / SELENE

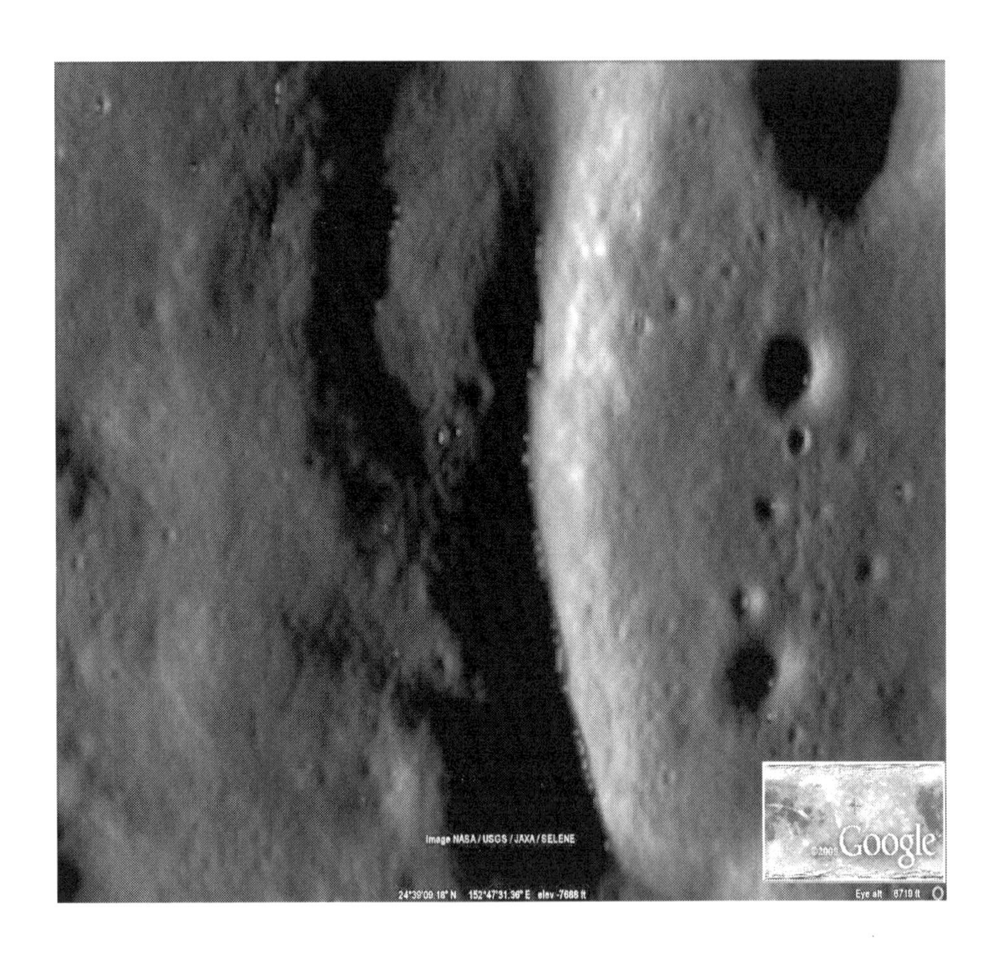
Image NASA / USGS / JAXA / SELENE
24°39'09.18" N 152°47'31.36" E elev -7688 ft
Google
Eye alt 6719 ft

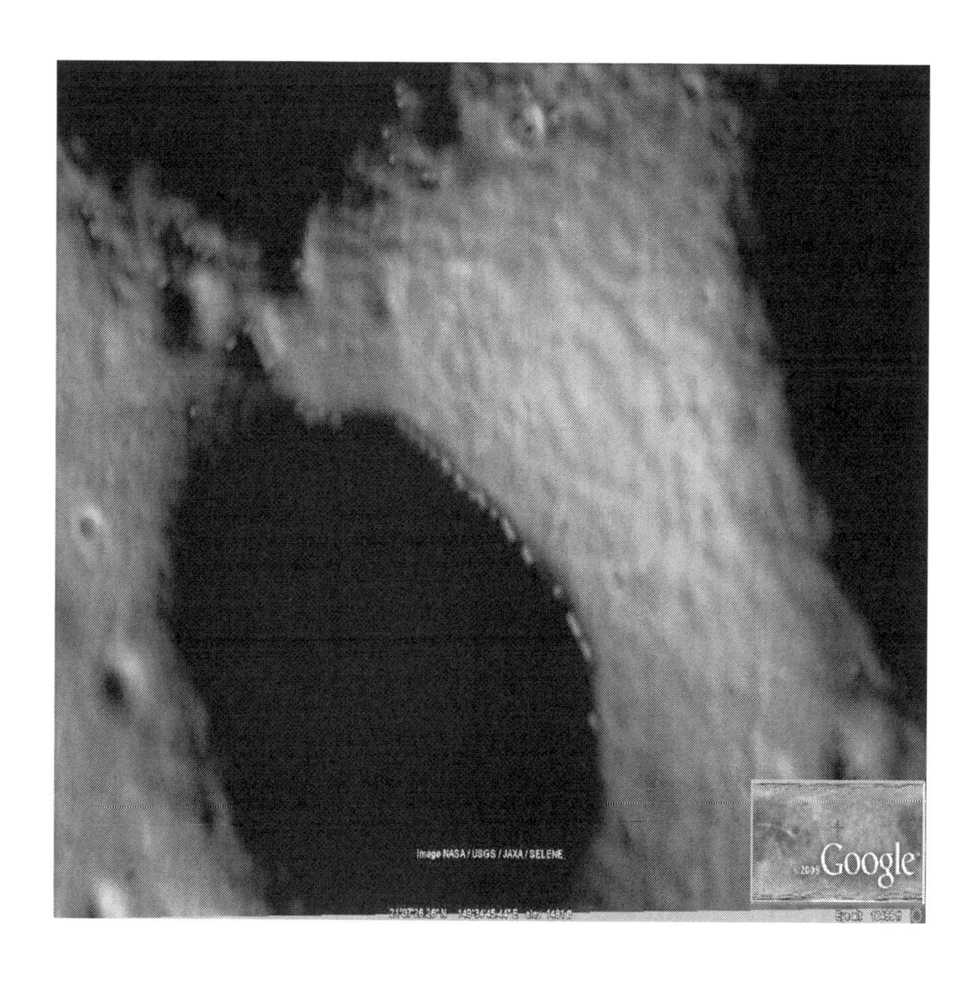
Image NASA / USGS / JAXA / SELENE
Google

Image NASA / USGS / JAXA / SELENE
Google
29°26'40.80" N 143°19'25.74" E elev -6231 ft
Eye alt 6409 ft

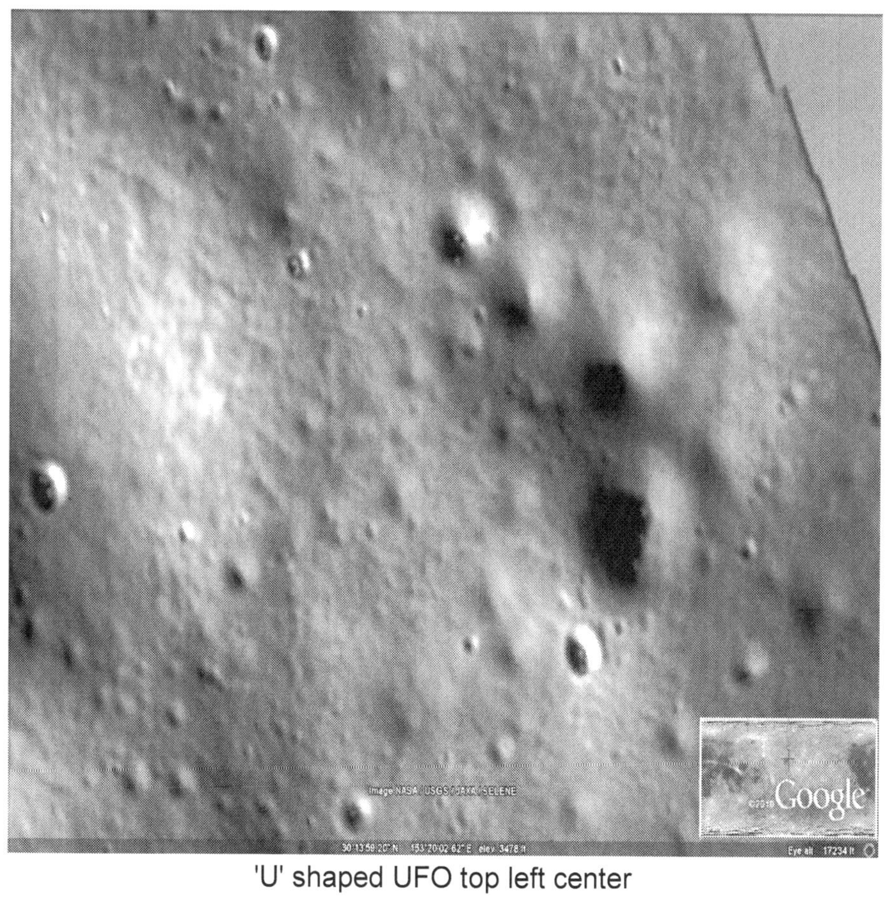

'U' shaped UFO top left center

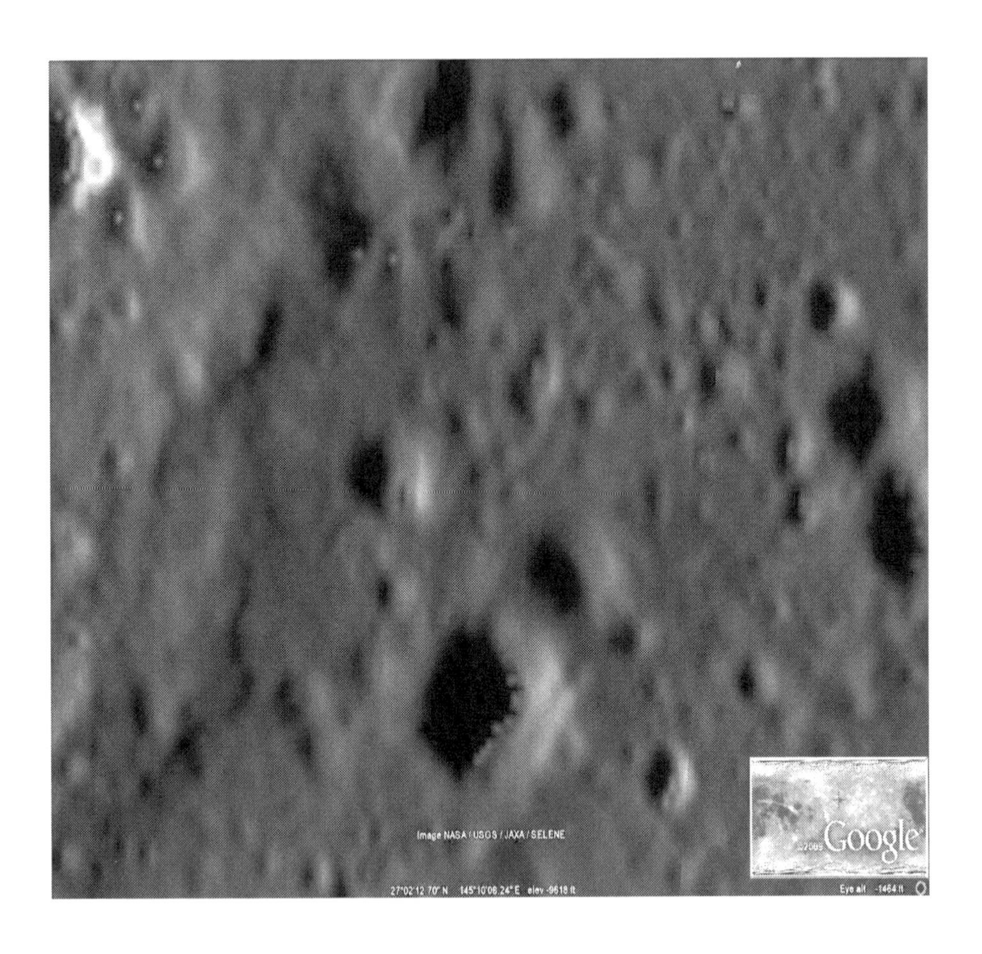
Image NASA / USGS / JAXA / SELENE
Google
27°02'12.70" N 145°10'08.24" E elev -9618 ft
Eye alt -1464 ft

Image NASA / USGS / JAXA / SELENE
Google
30°51'58.81" N 142°51'27.95" E elev 2382 ft
Eye alt 15236 ft

Image NASA / USGS / JAXA / SELENE
Google
29°25'18.51" N 143°18'55.26" E elev -6237 ft
Eye alt 5008 ft

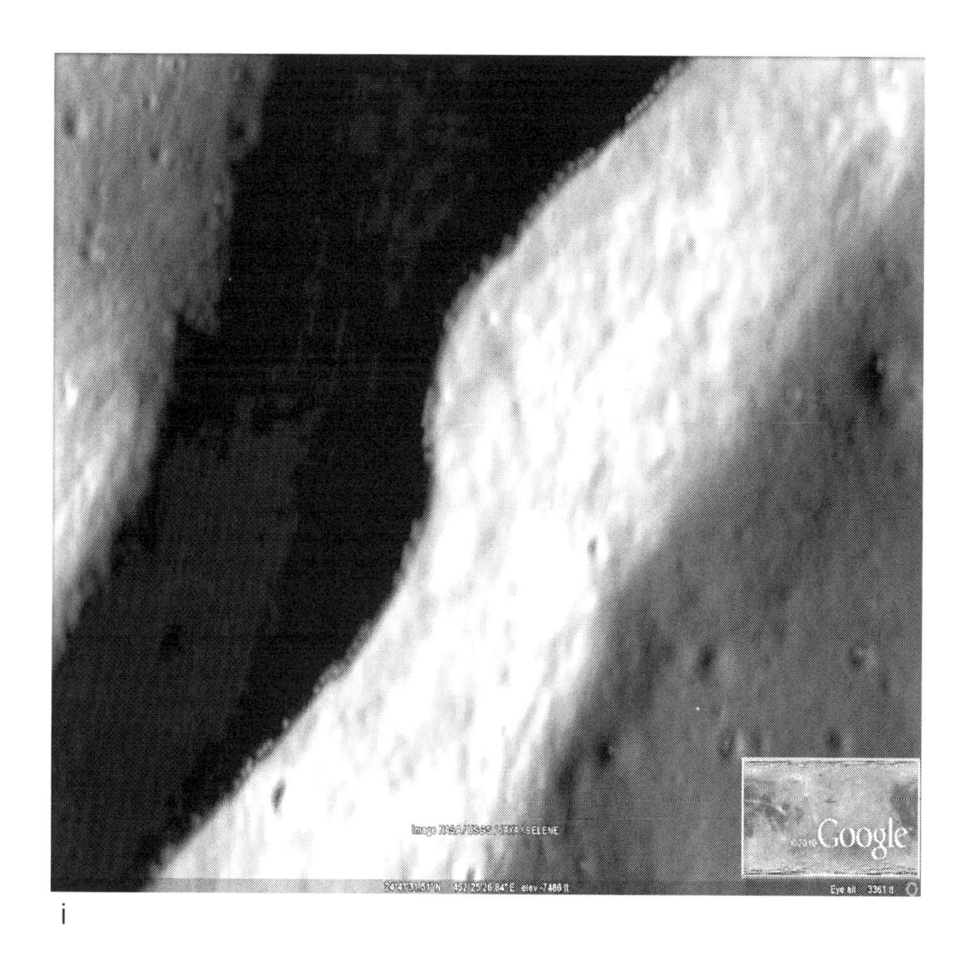

i

Image NASA / USGS / JAXA / SELENE
Google
29°54'37.08" N 145°23'18.43" E elev -5390 ft
Eye alt 14556 ft

Image NASA / USGS / JAXA / SELENE
Google
143°13'41.04" E elev -5295 ft
Eye alt 8855 ft

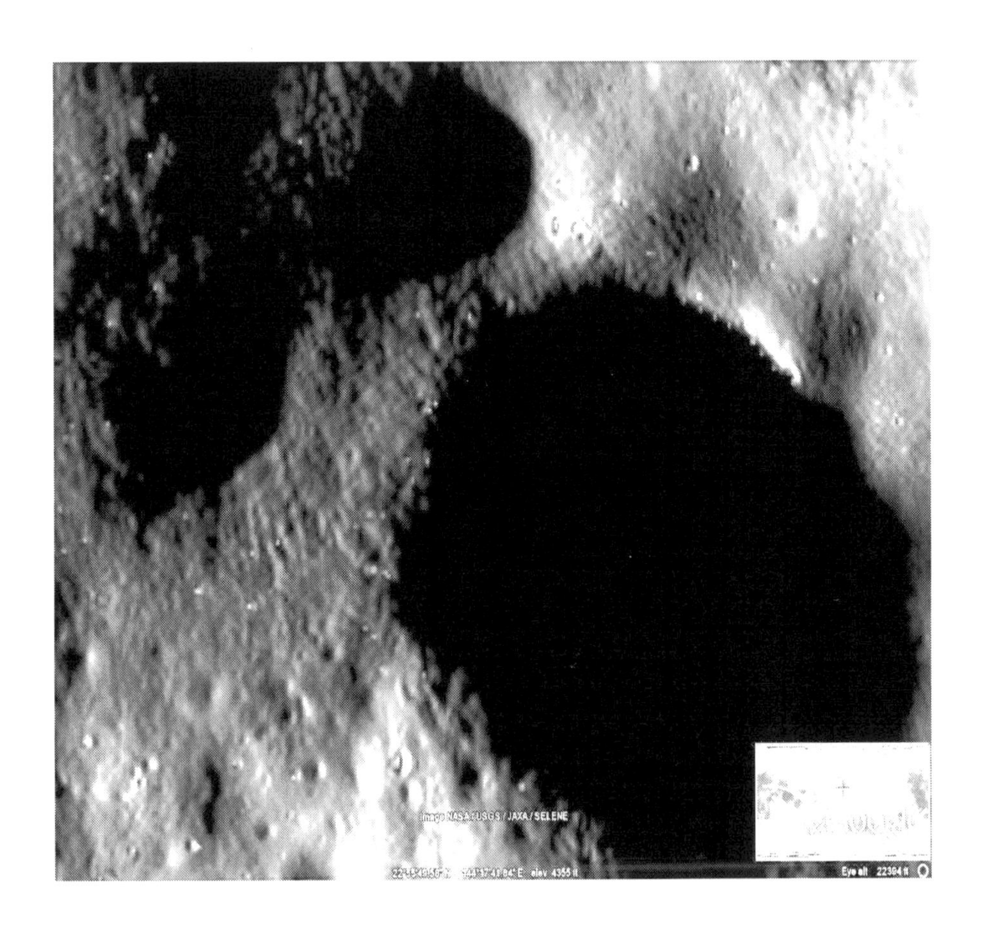
Image NASA / USGS / JAXA / SELENE
elev 4355 ft
Eye alt 22394 ft

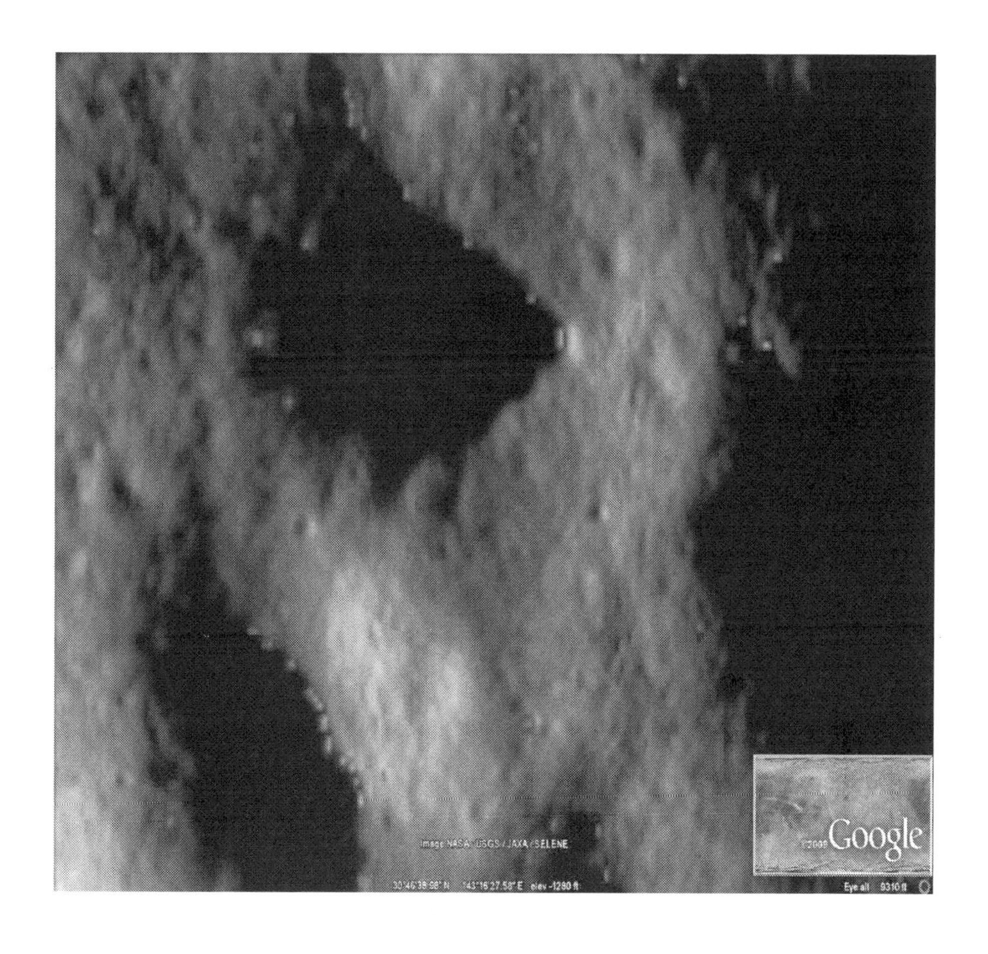
Google
Eye alt 9310 ft

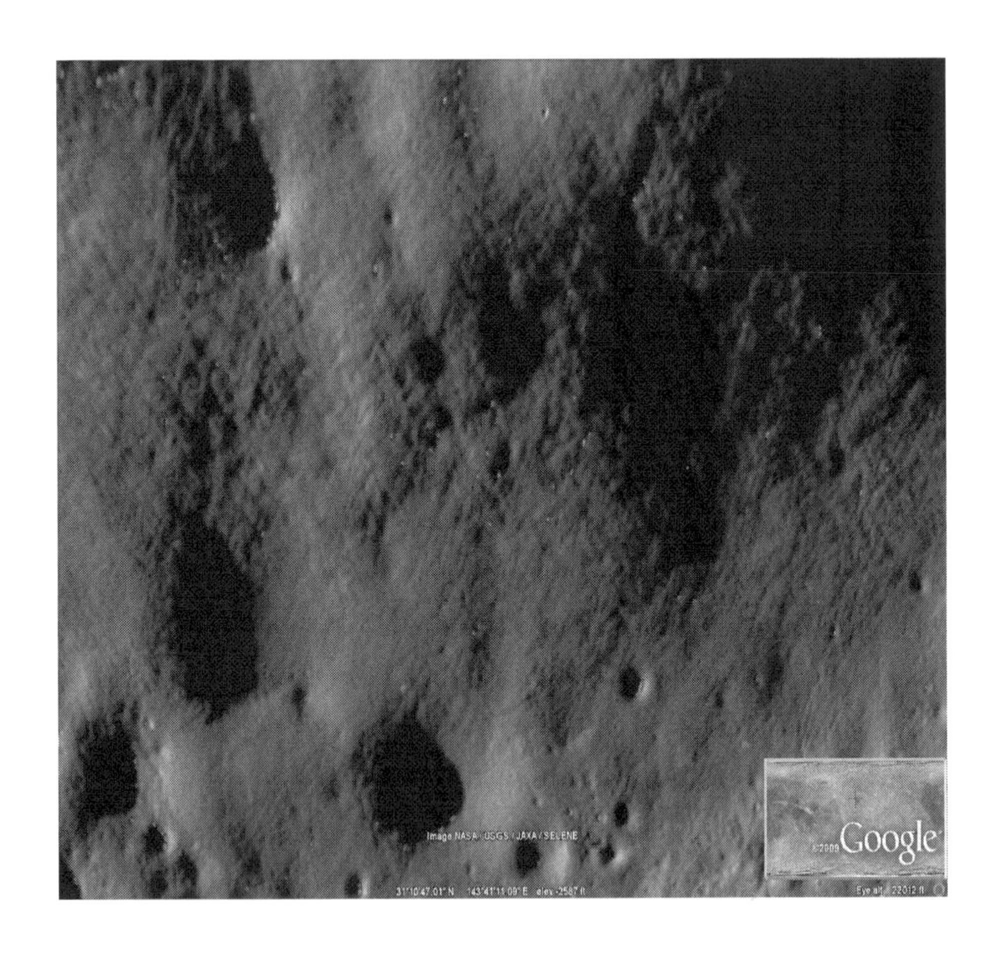
Image NASA / USGS / JAXA / SELENE
Google
31°10'47.01" N 143°41'11.09" E elev -2587 ft
Eye alt 22012 ft

Image NASA / USGS / JAXA / SELENE
Google

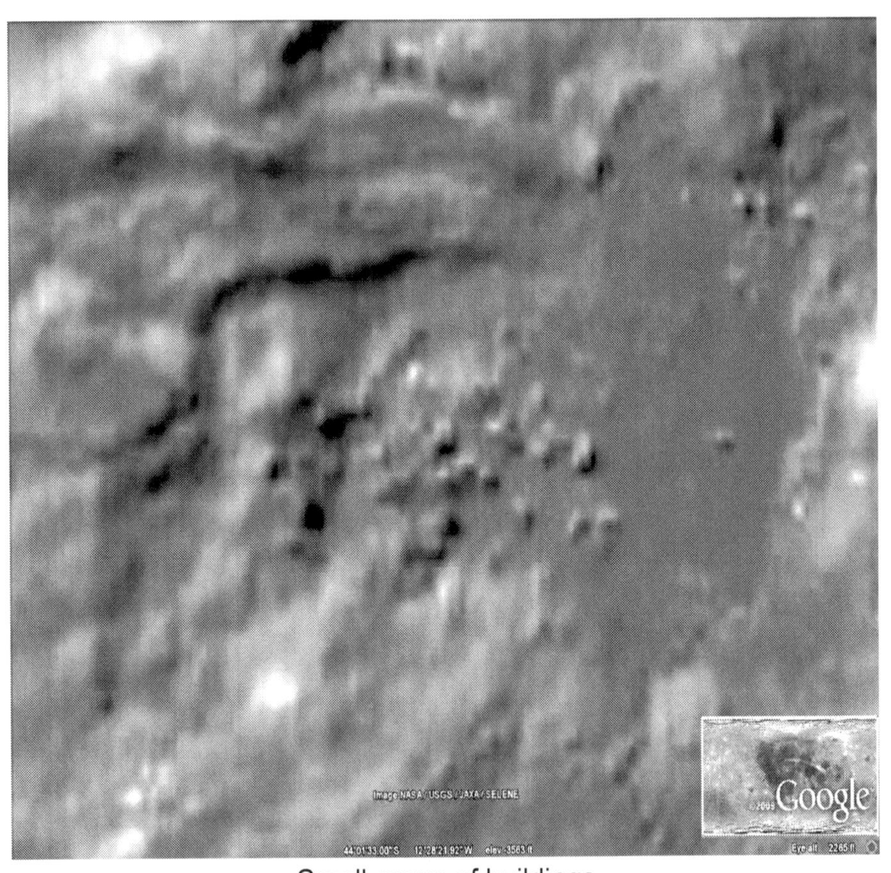

Small group of buildings

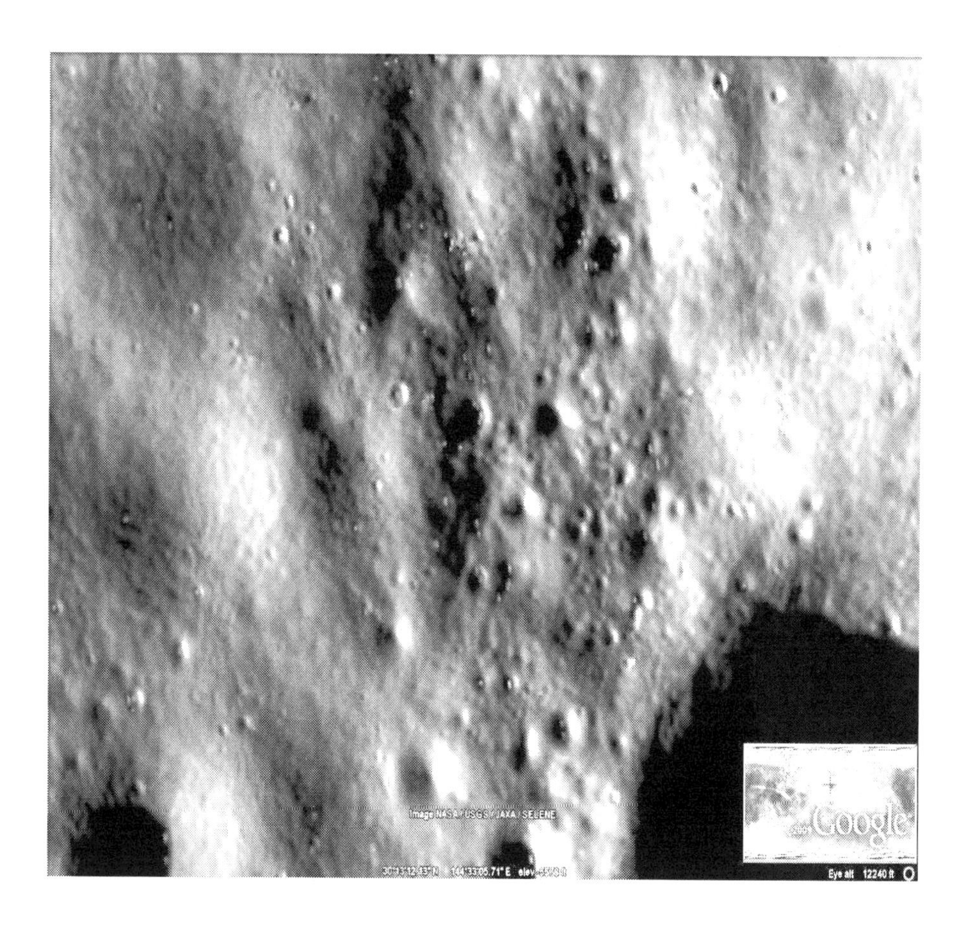
Image NASA / USGS / JAXA / SELENE
144°33'05.71" E
Google
Eye alt 12240 ft

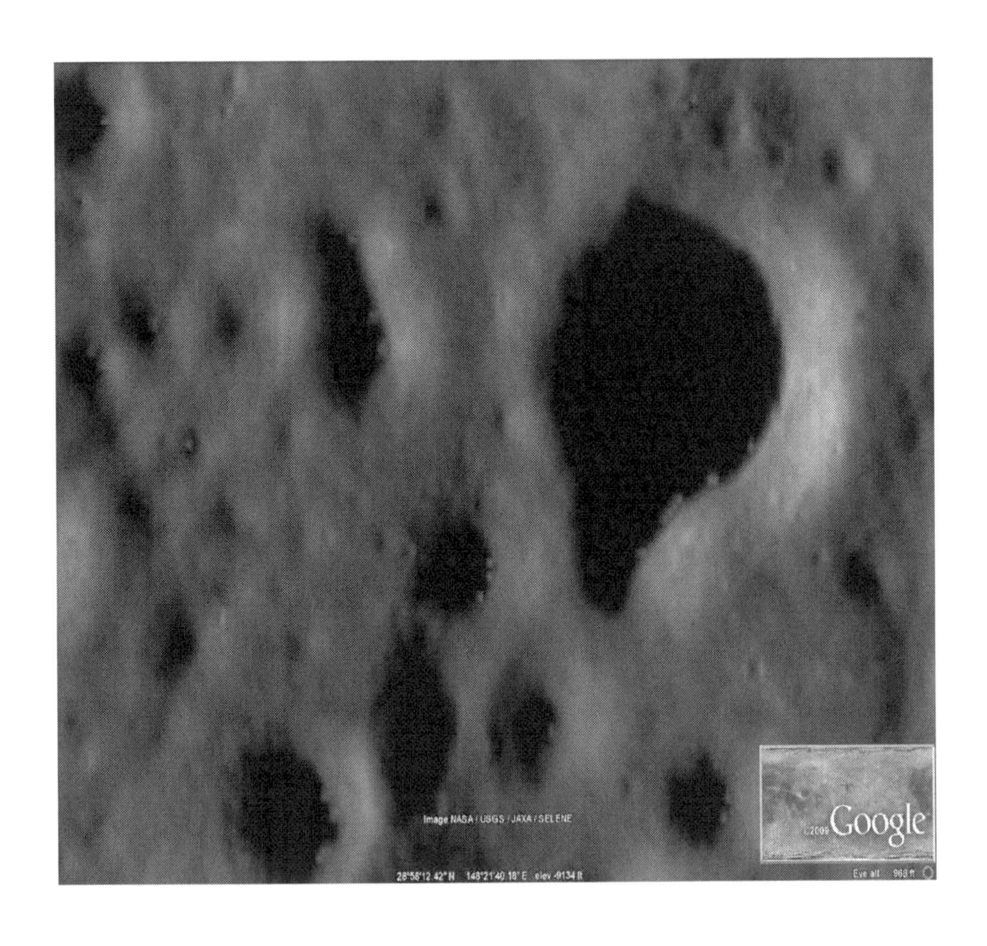

Image NASA / USGS / JAXA / SELENE
Google

Image NASA / USGS / JAXA / SELENE
Google
30°39'26.08" N 143°25'42.45" E elev -3695 ft

Image NASA / USGS / JAXA / SELENE
Google
28°53'38.33" N 148°08'29.81" E elev -9290 ft
Eye alt 6553 ft

Several cigar shaped UFO, look how they all face the same direction

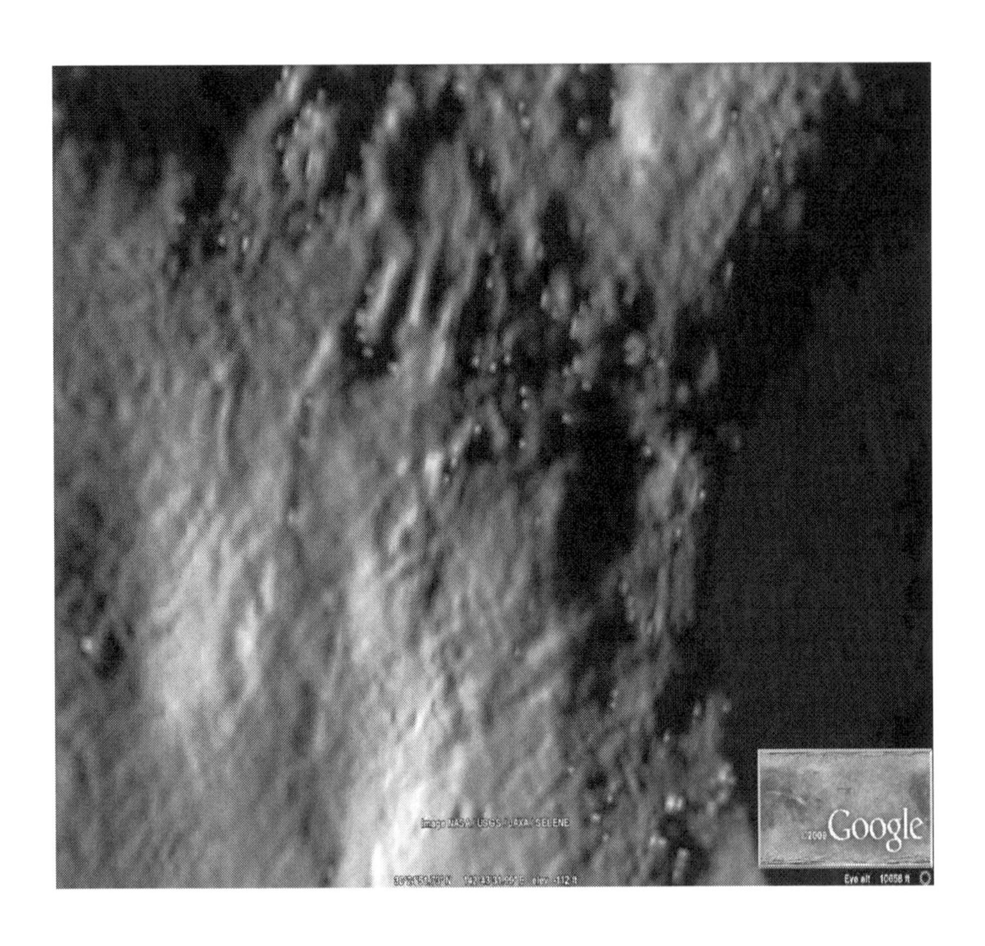
Image NASA / USGS / JAXA / SELENE
Google

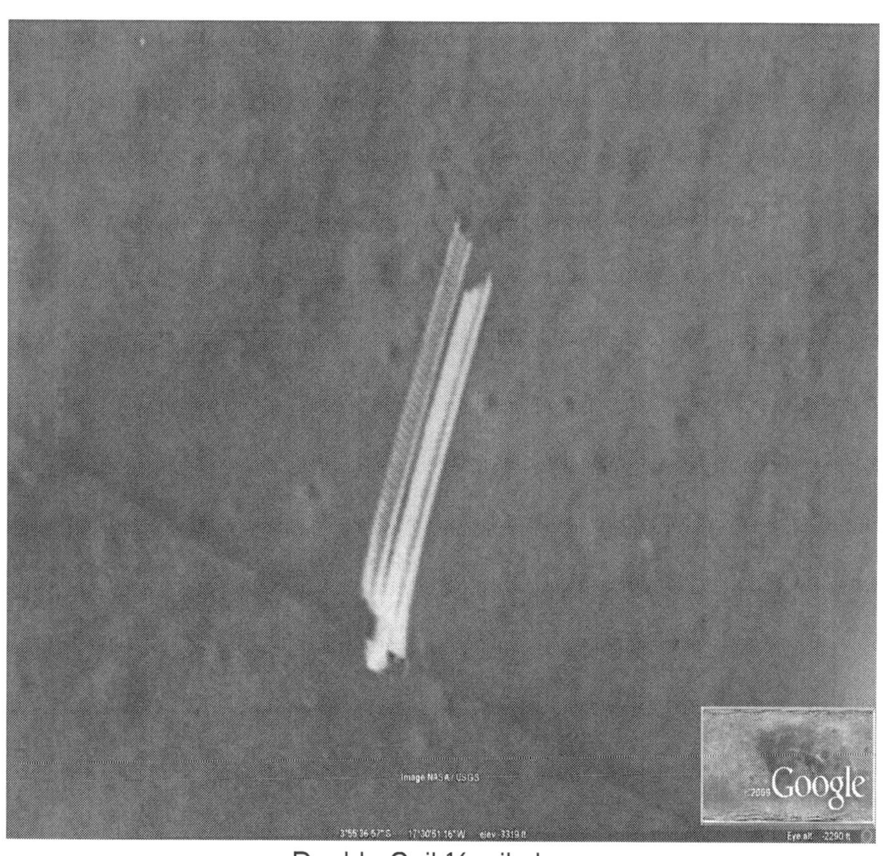

Double Coil ¼ mile long

Image NASA / USGS / JAXA / SELENE
Google
143°12'14.27" E elev -5535 ft

Dec 31, 2003 3:31 am
Image NASA / USGS / JAXA / SELENE
Google

Dec 31, 2003 1:16 am
Dec 30, 2003
Dec 31, 2003
Untitled Placemark
Image NASA / USGS / JAXA / SELENE
Google
145°06'36.50"E elev -4778 ft
Eye alt 6420 ft

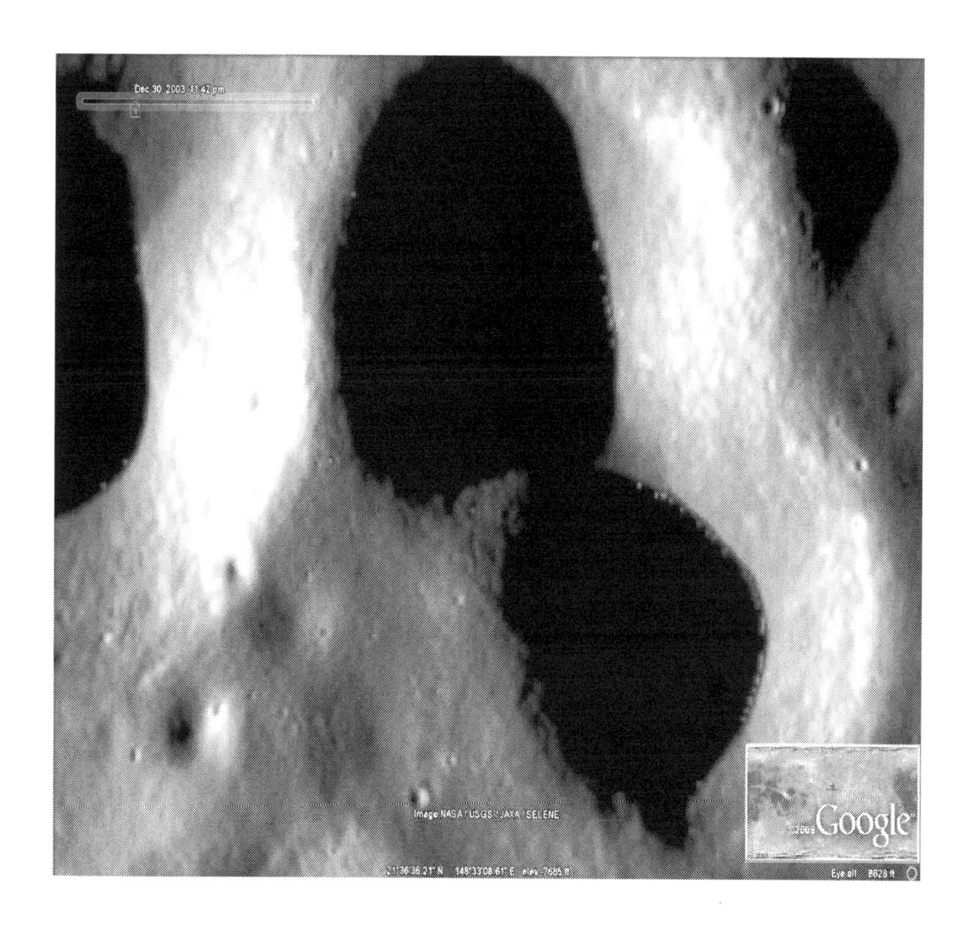
Dec 30 2003 11:42 pm
Image NASA / USGS / JAXA / SELENE
Google
21°36'36.21" N 148°33'08.61" E elev -7685 ft
Eye alt 8628 ft

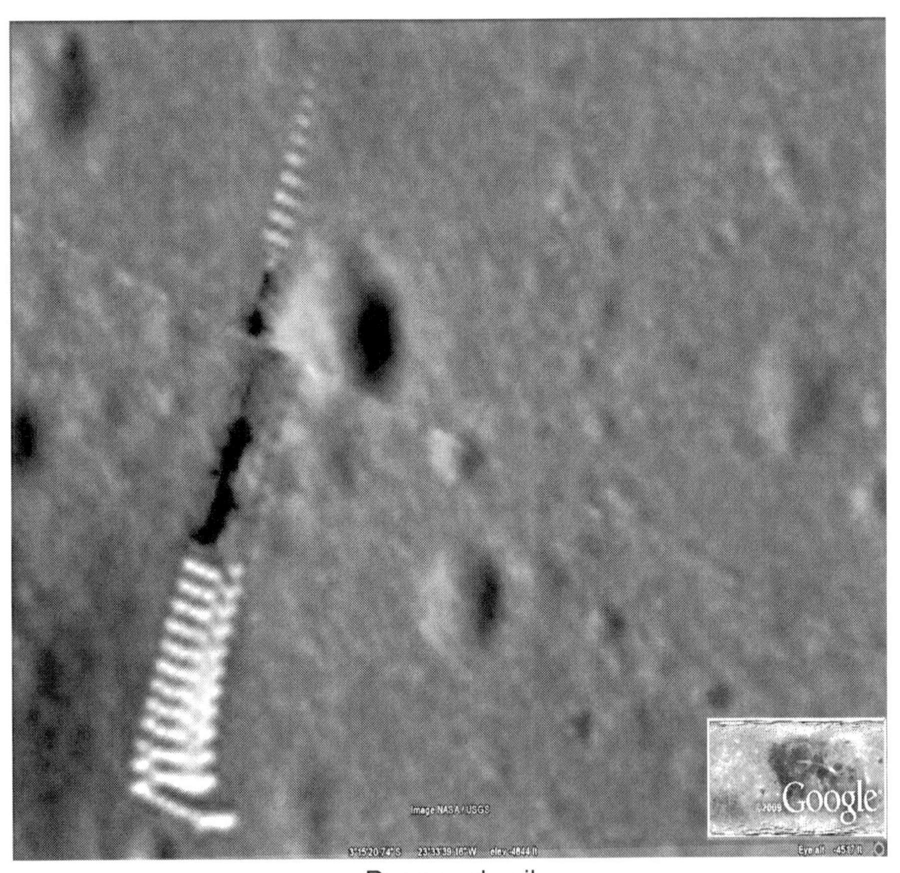

Damaged coil

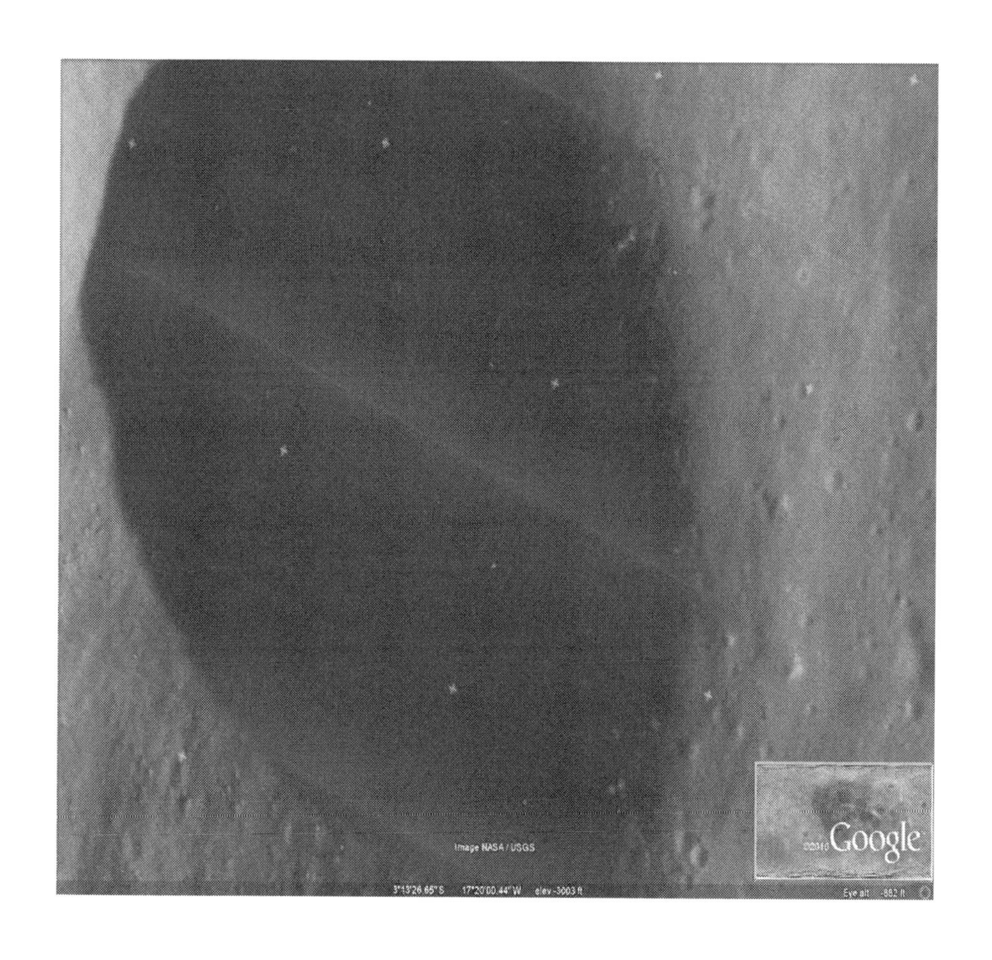
Image NASA / USGS
Google
3°13'26.65"S 17°20'00.44" W elev -3003 ft
Eye alt

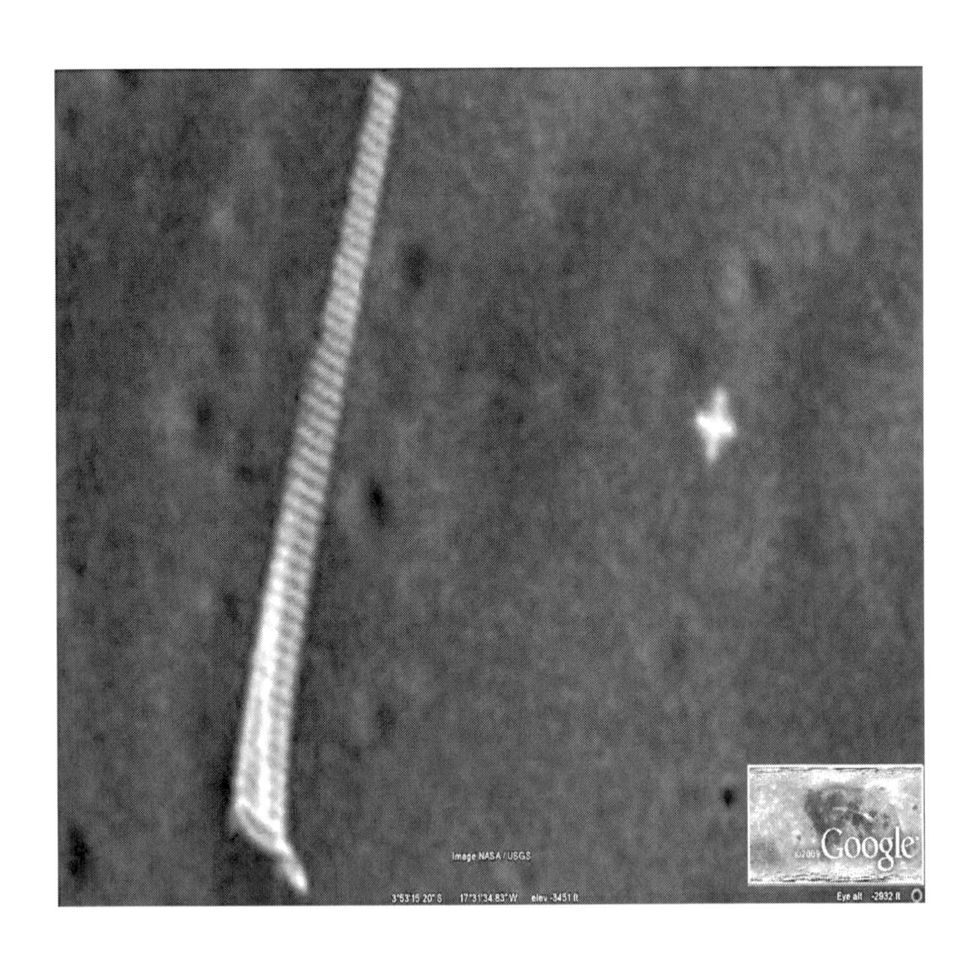
Image NASA / USGS
Google
3°53'15.20" S
17°31'34.83" W
elev -3451 ft
Eye alt -2932 ft

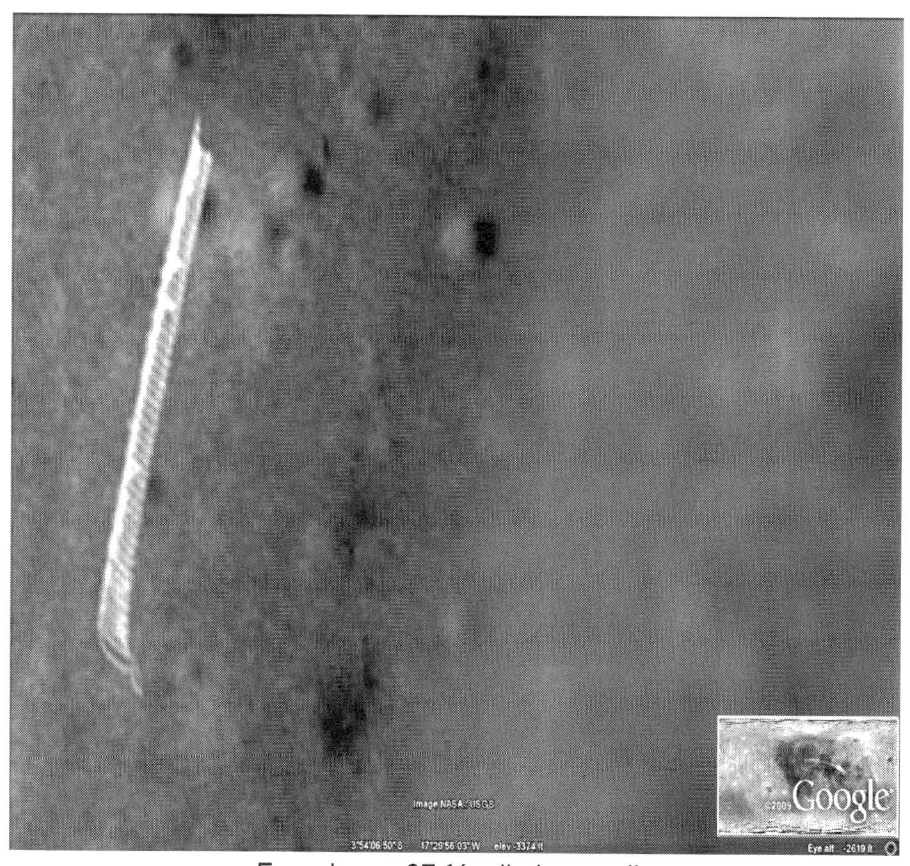

Found over 37 ¼ mile long coils

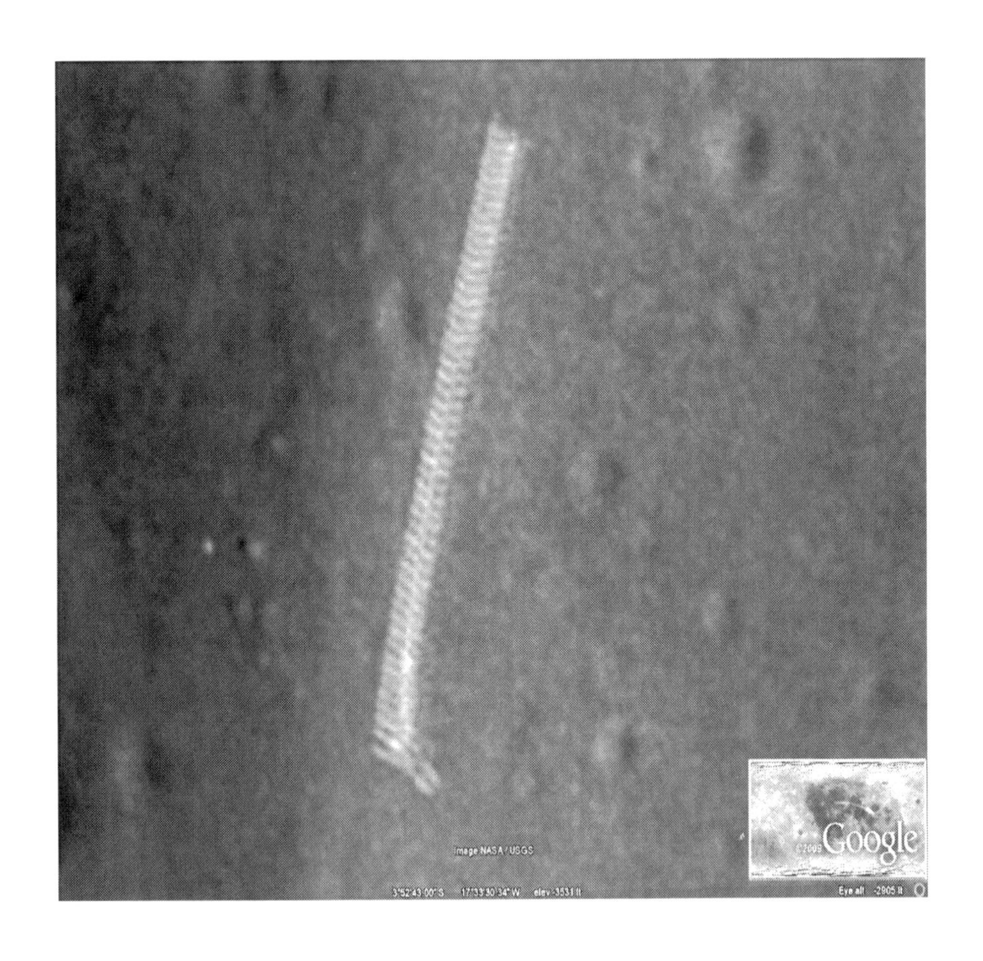
Image NASA / USGS
©2009 Google
3°52'43.00"S 17°33'30.34"W elev -3531 ft
Eye alt -2905 ft

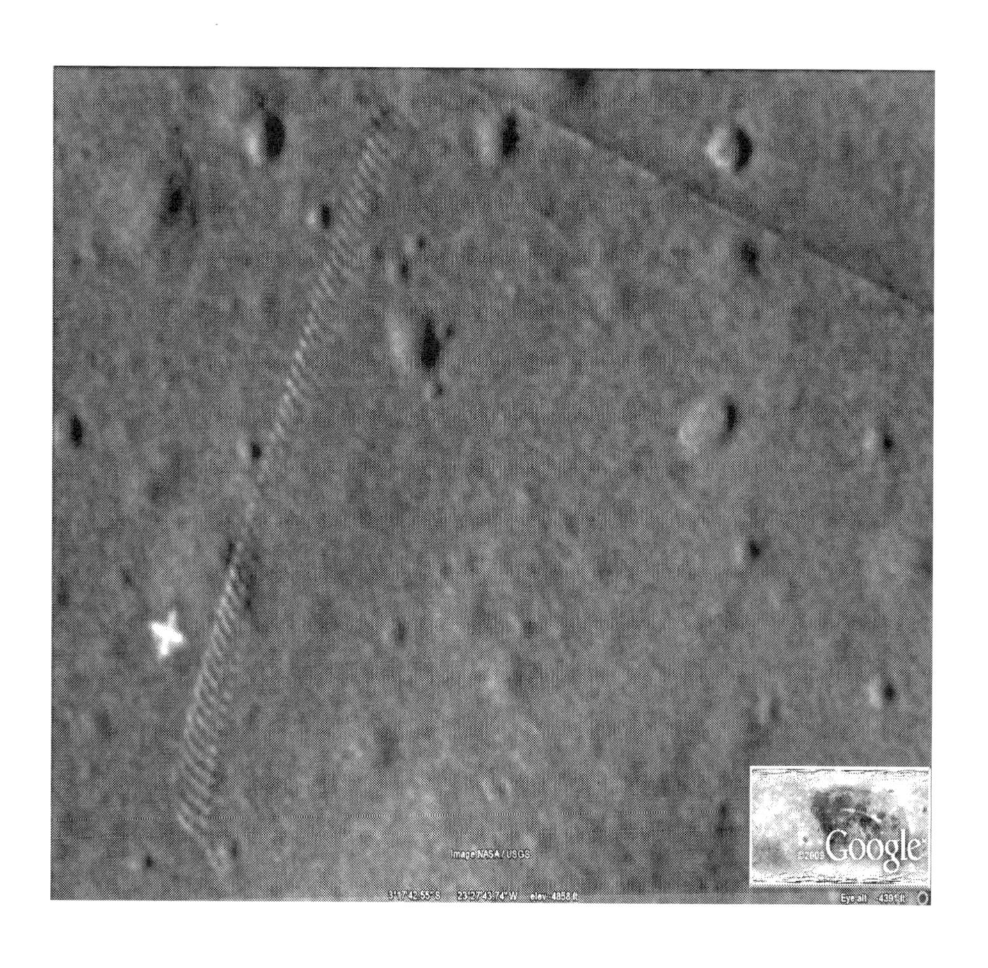
Image NASA / USGS
©2009 Google
3°17'42.55" S
23°27'43.74" W
elev -4858 ft
Eye alt

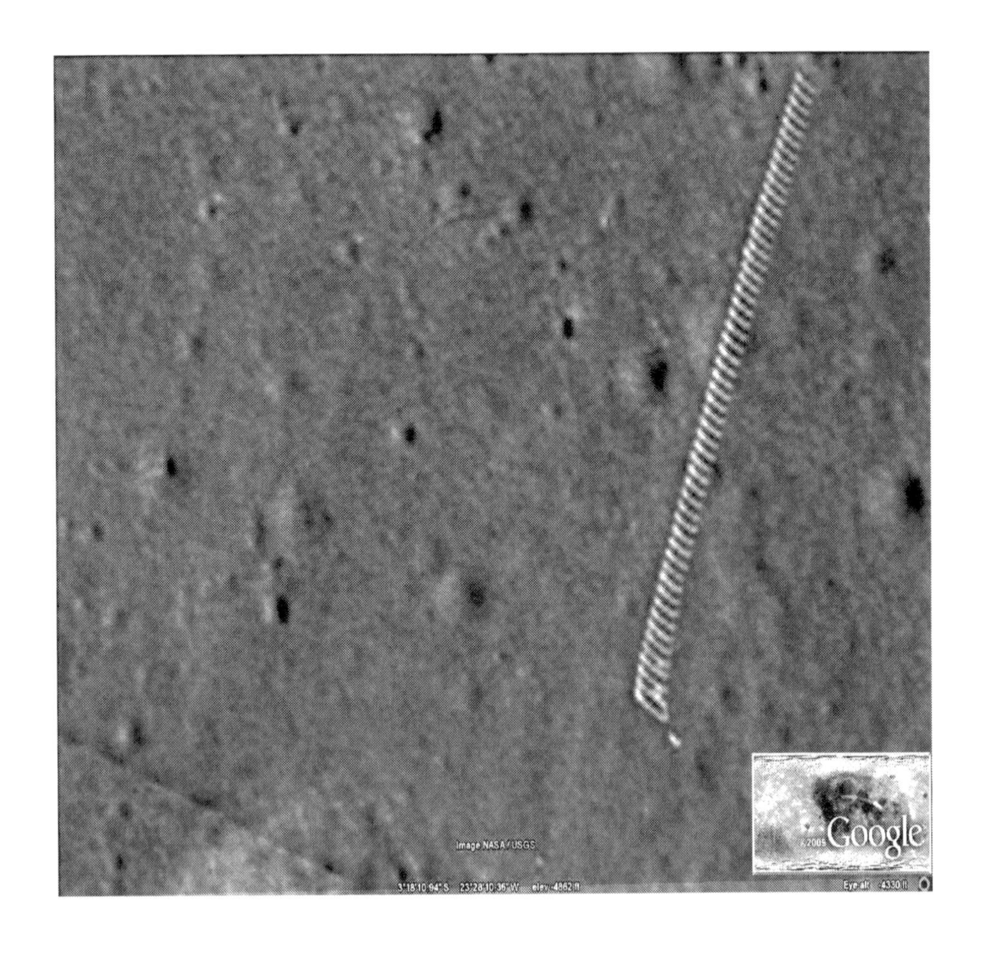
Image NASA / USGS
Google
3°18'10.94" S 23°28'10.36" W elev -4862 ft
Eye alt 4330 ft

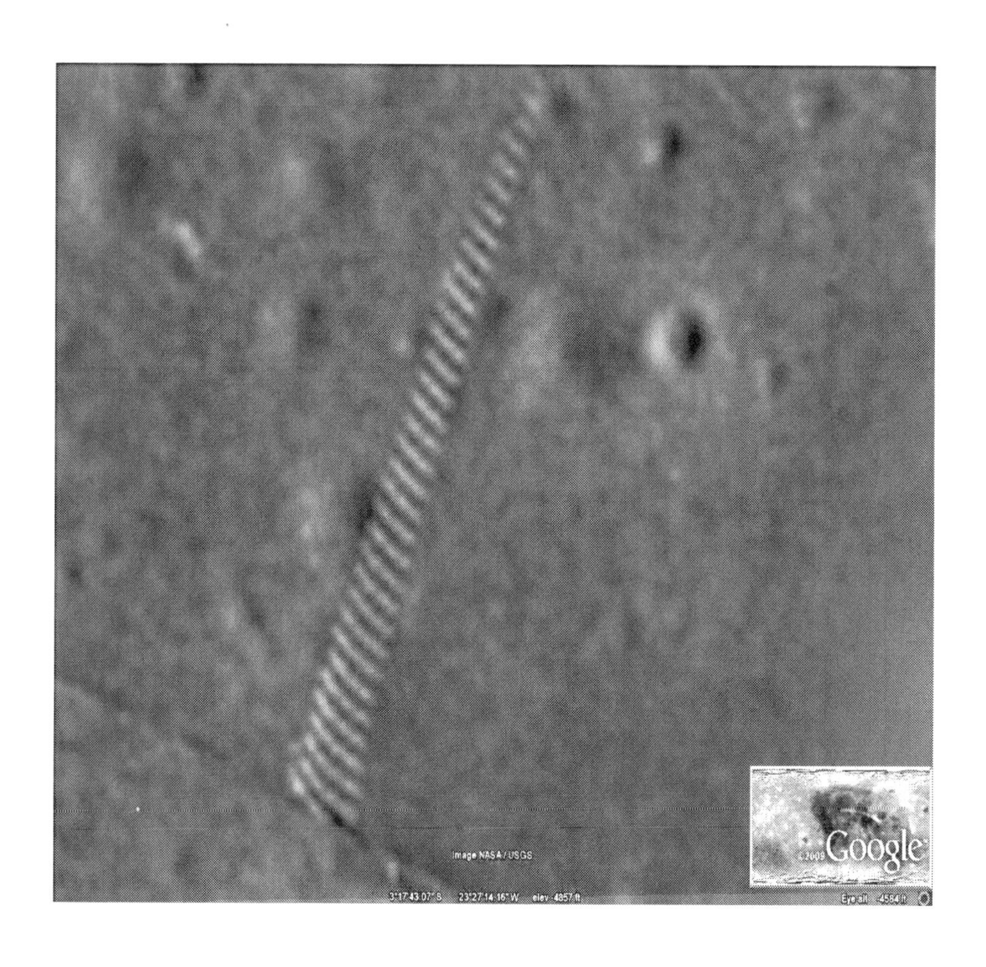
Image NASA / USGS
Google
3°17'43.07" S 23°27'14.16" W elev -4857 ft
Eye alt -4584 ft

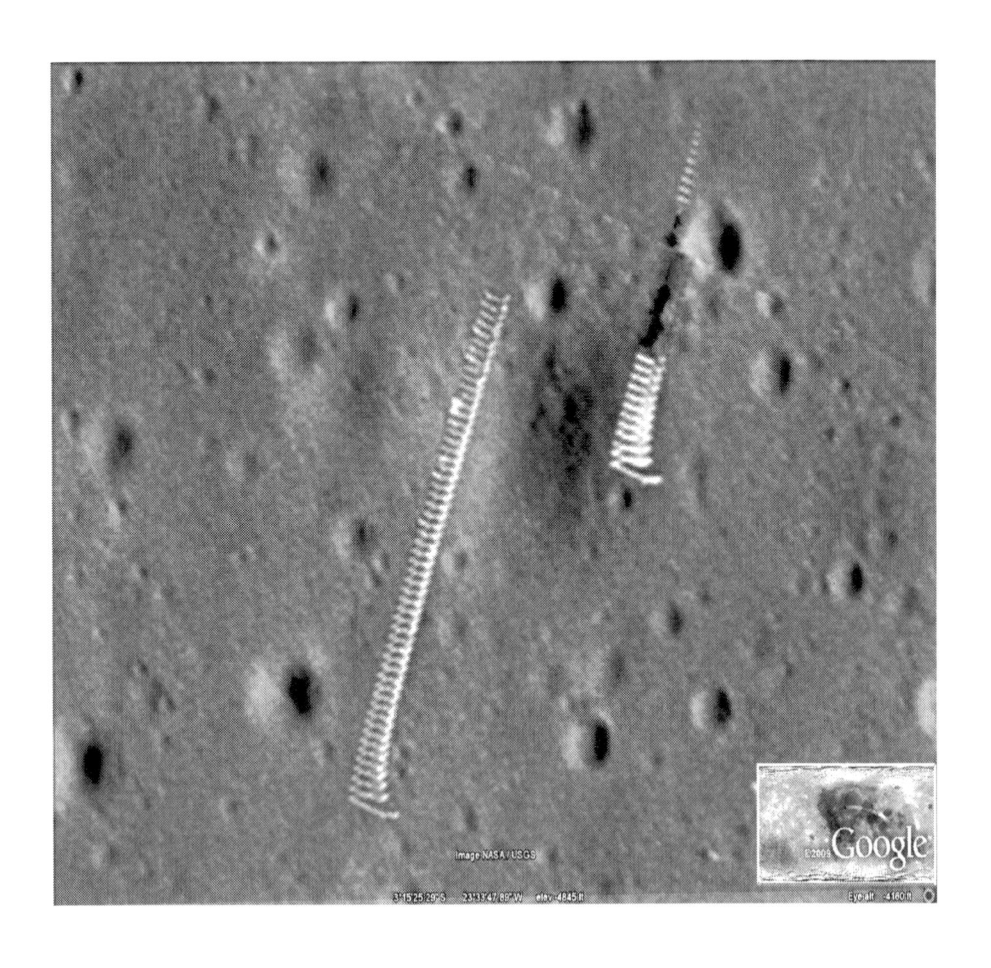
Image NASA / USGS
©2009 Google
3°15'25.29" S   23°33'47.89" W   elev -4845 ft
Eye alt -4160 ft

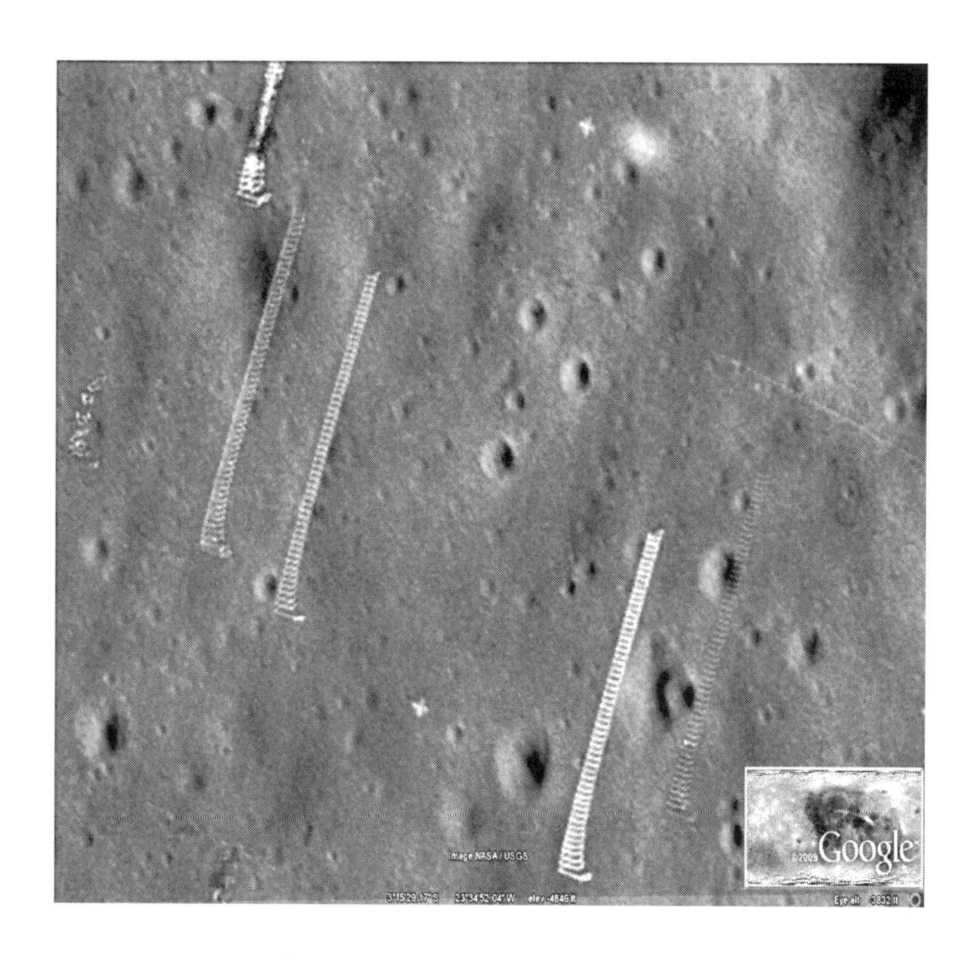
Image NASA / USGS
©2009 Google
3°15'29.17"S 23°34'52.04"W elev -4846 ft
Eye alt 3832 ft

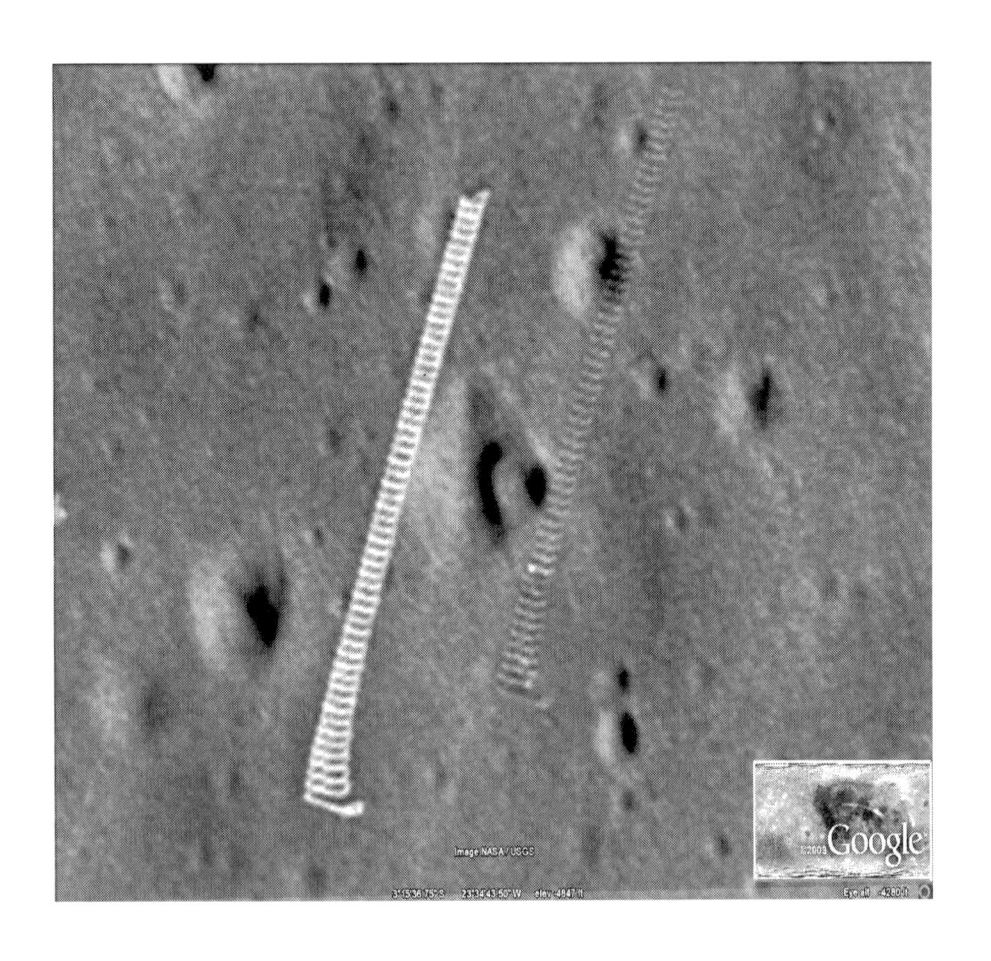
Image NASA / USGS
Google

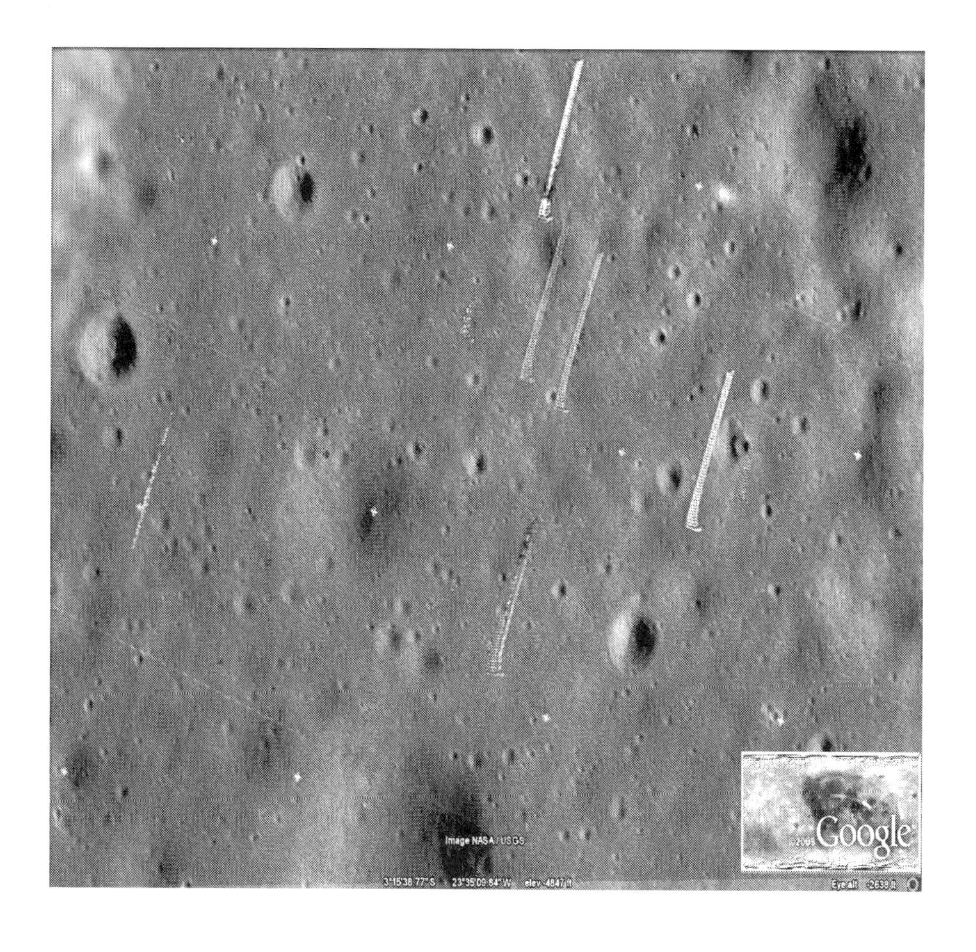
Image NASA / USGS
Google
3°15'38.77" S
23°35'09.84" W
elev -4847 ft
Eye alt
2639 ft

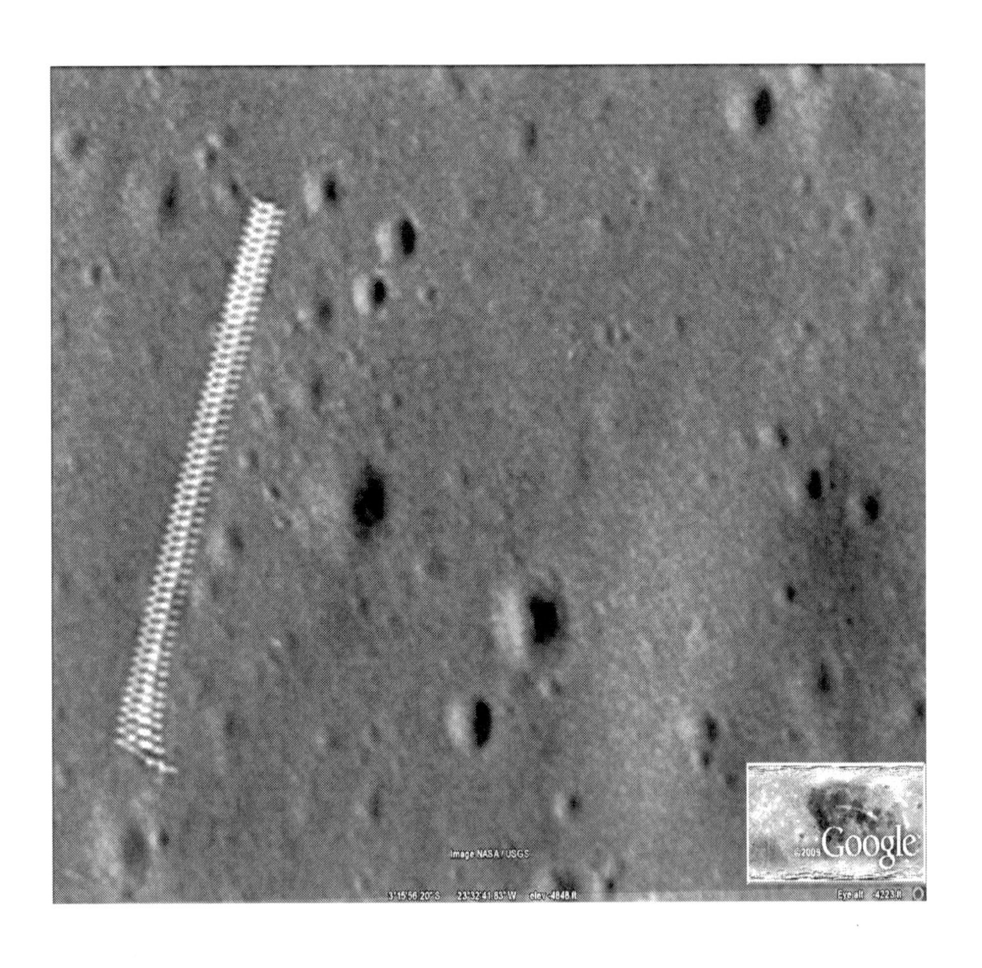
Image NASA / USGS
©2009 Google
3°15'56.20" S 23°32'41.83" W elev -4848 ft
Eye alt -4223 ft

Image NASA/USGS
Google
3°17'43.66" S 23°27'49.08" W elev -4858 ft
Eye alt -4318 ft

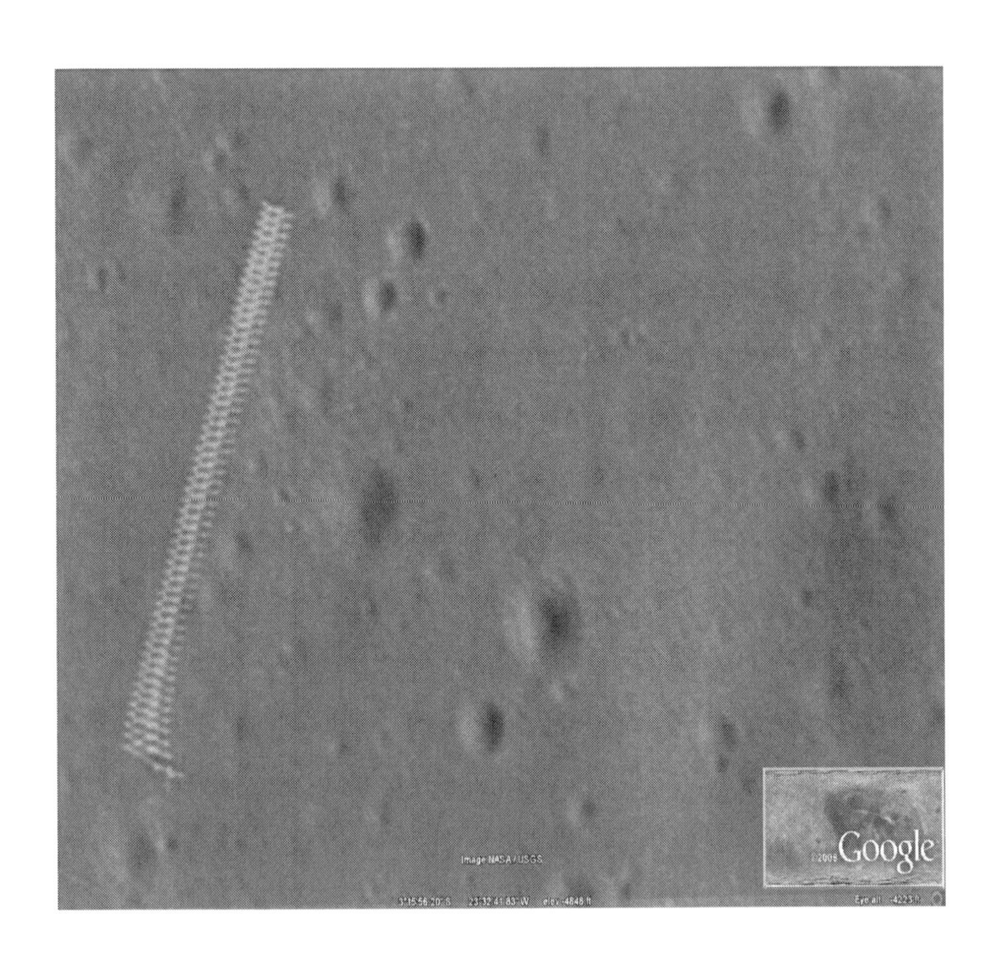
Image NASA / USGS
Google

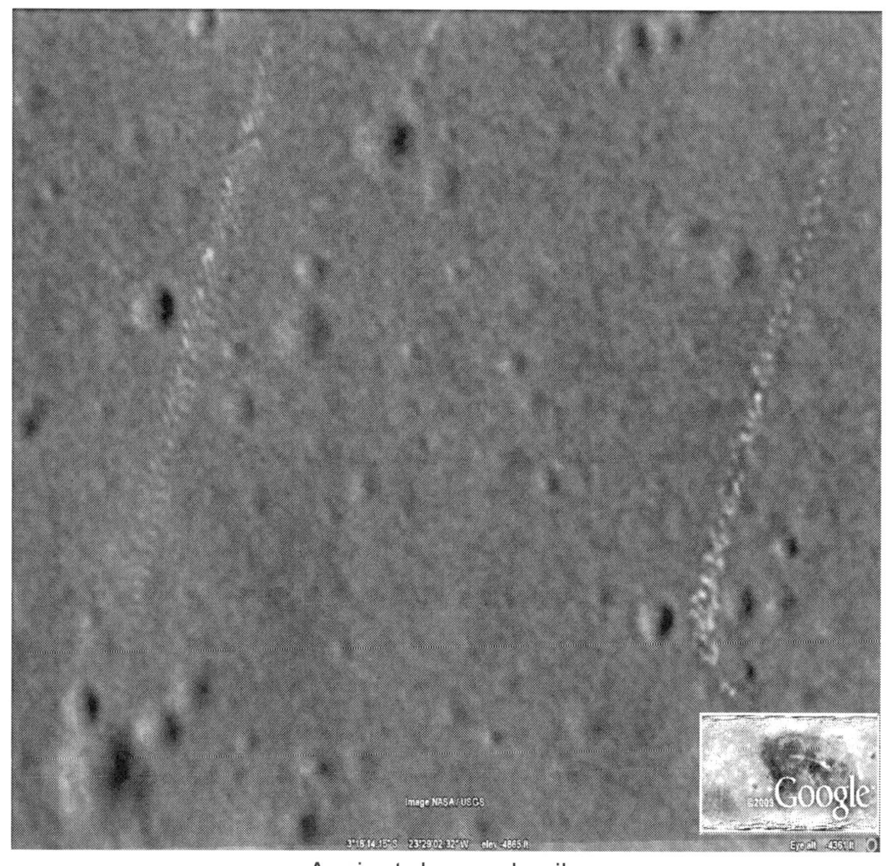

Ancient decayed coils

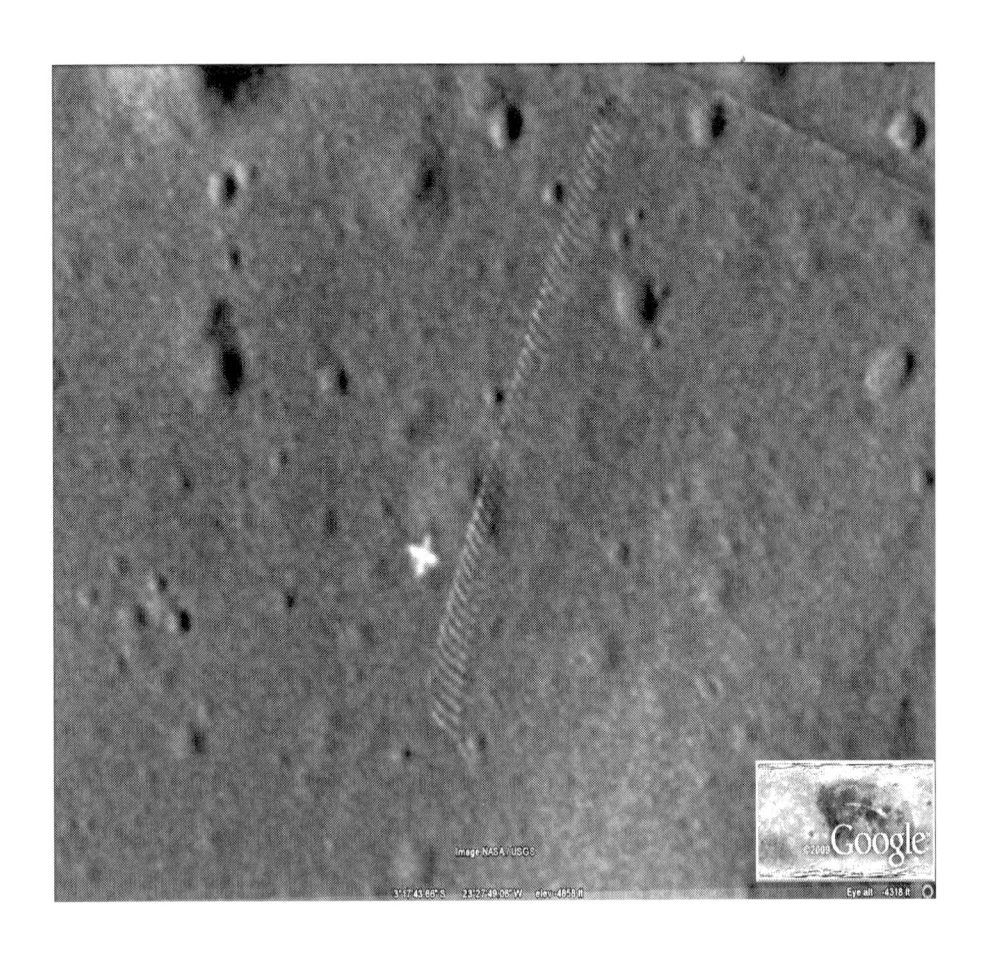
Image NASA / USGS
©2009 Google
3°17'43.66" S 23°27'49.08" W elev -4858 ft
Eye alt -4318 ft

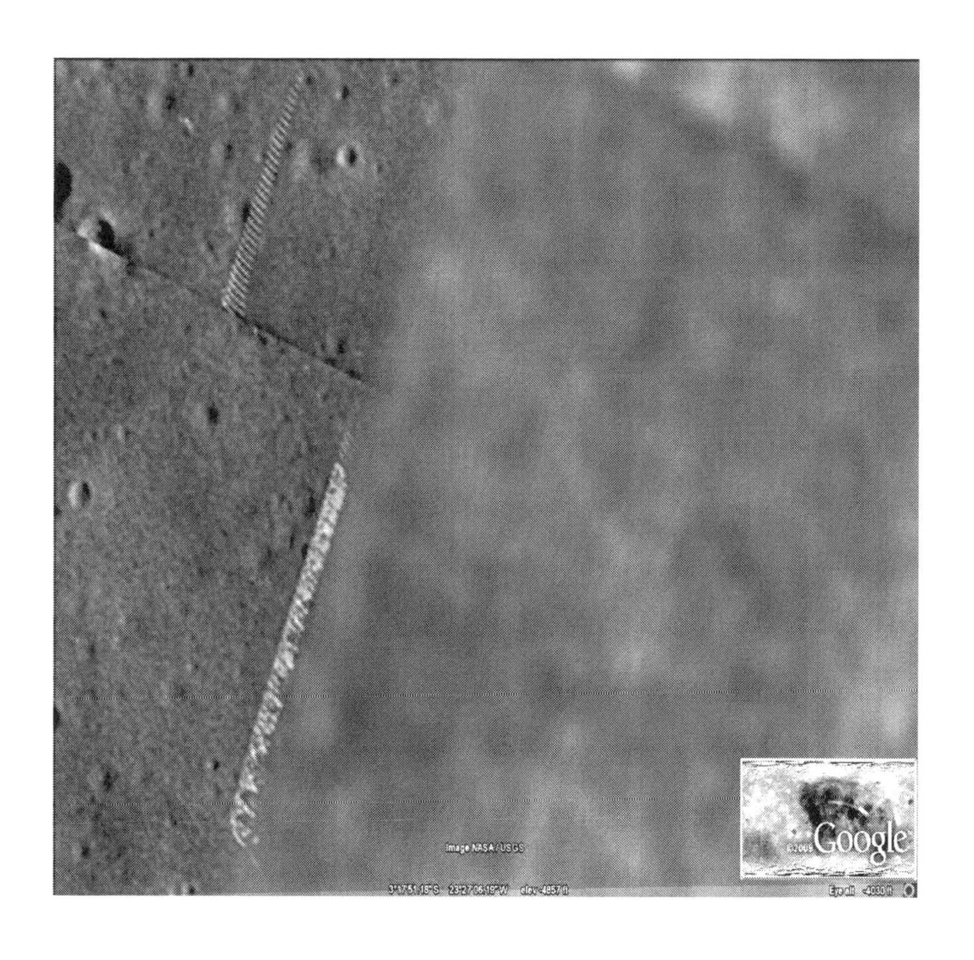

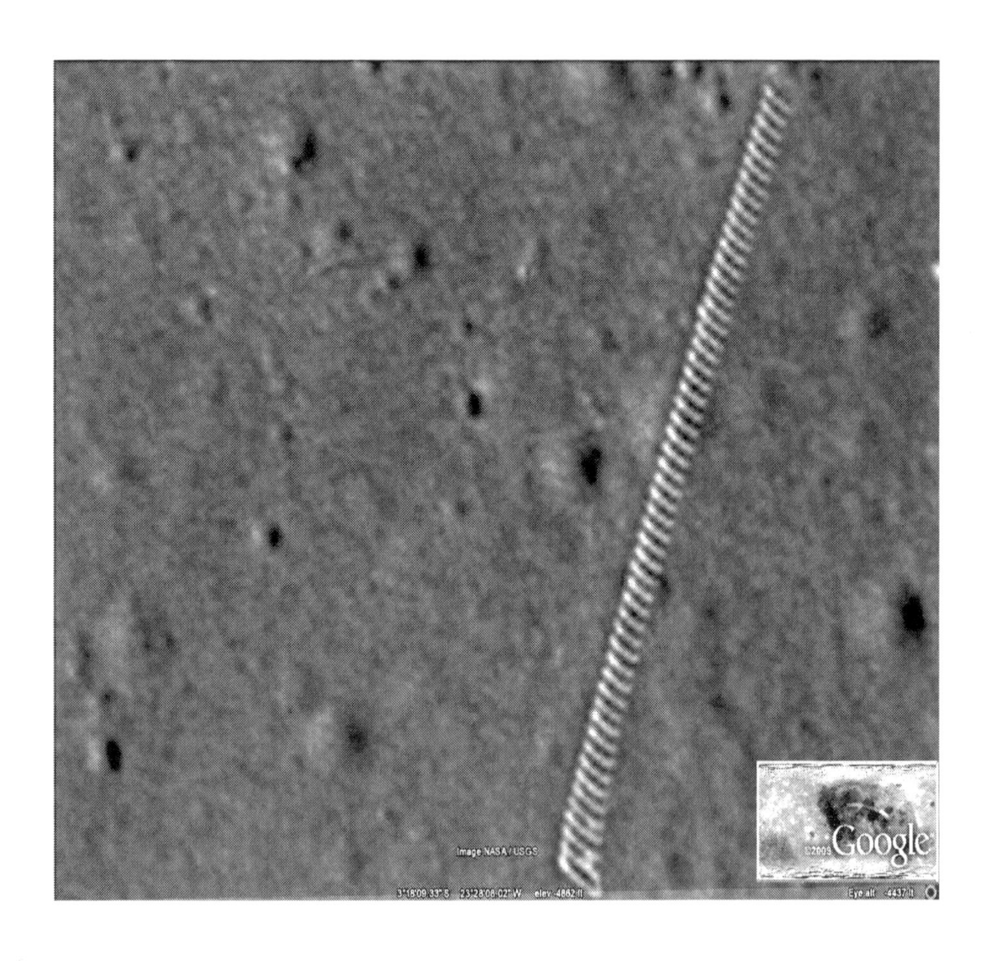
Image NASA / USGS
©2009 Google
3°18'09.33" S 23°28'08.02" W elev -4862 ft
Eye alt -4437 ft

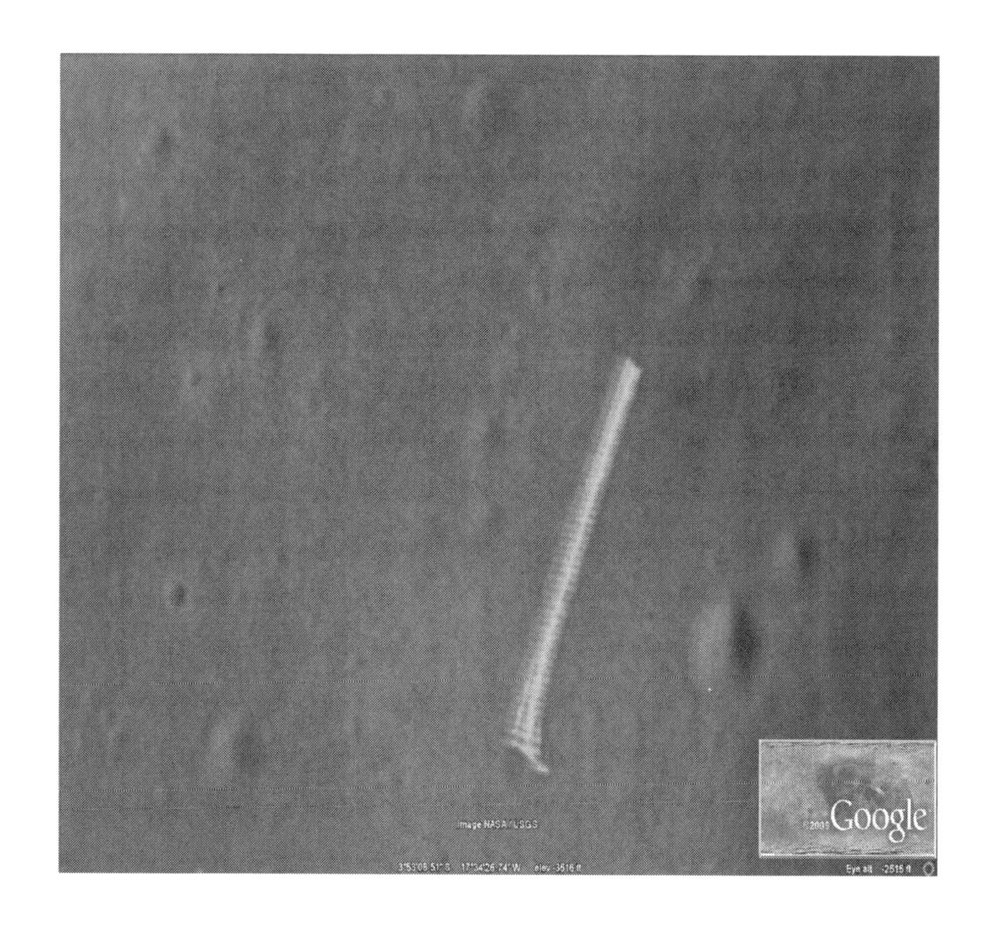

Image NASA / USGS
Google

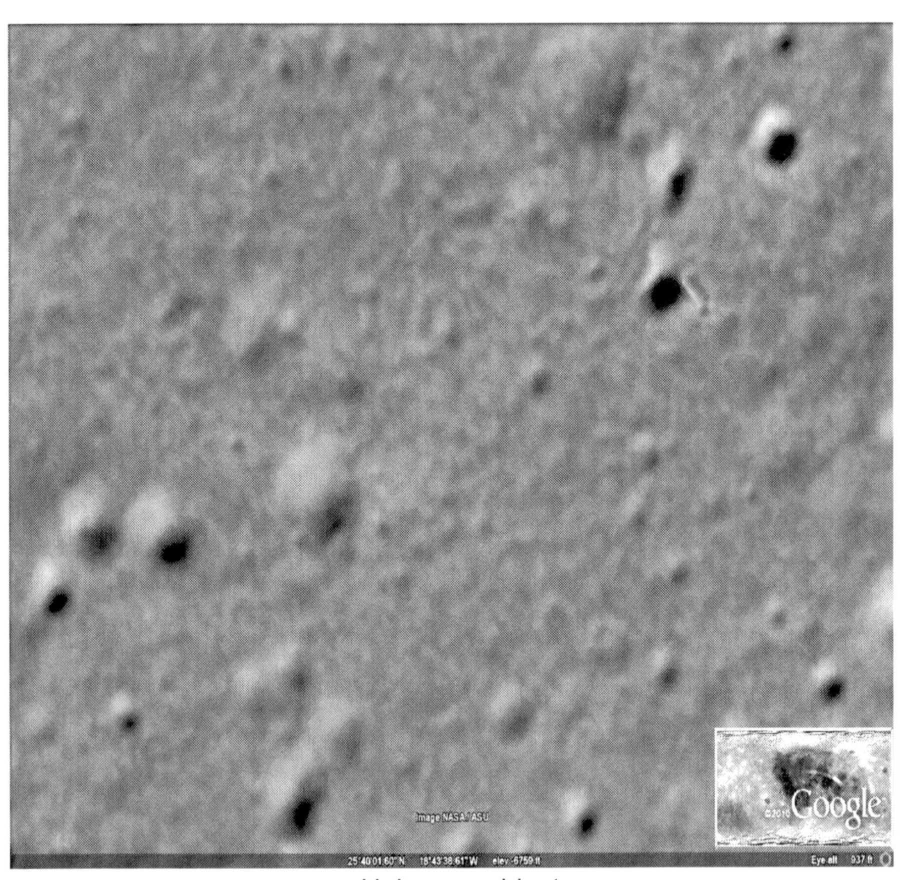

Unknown object

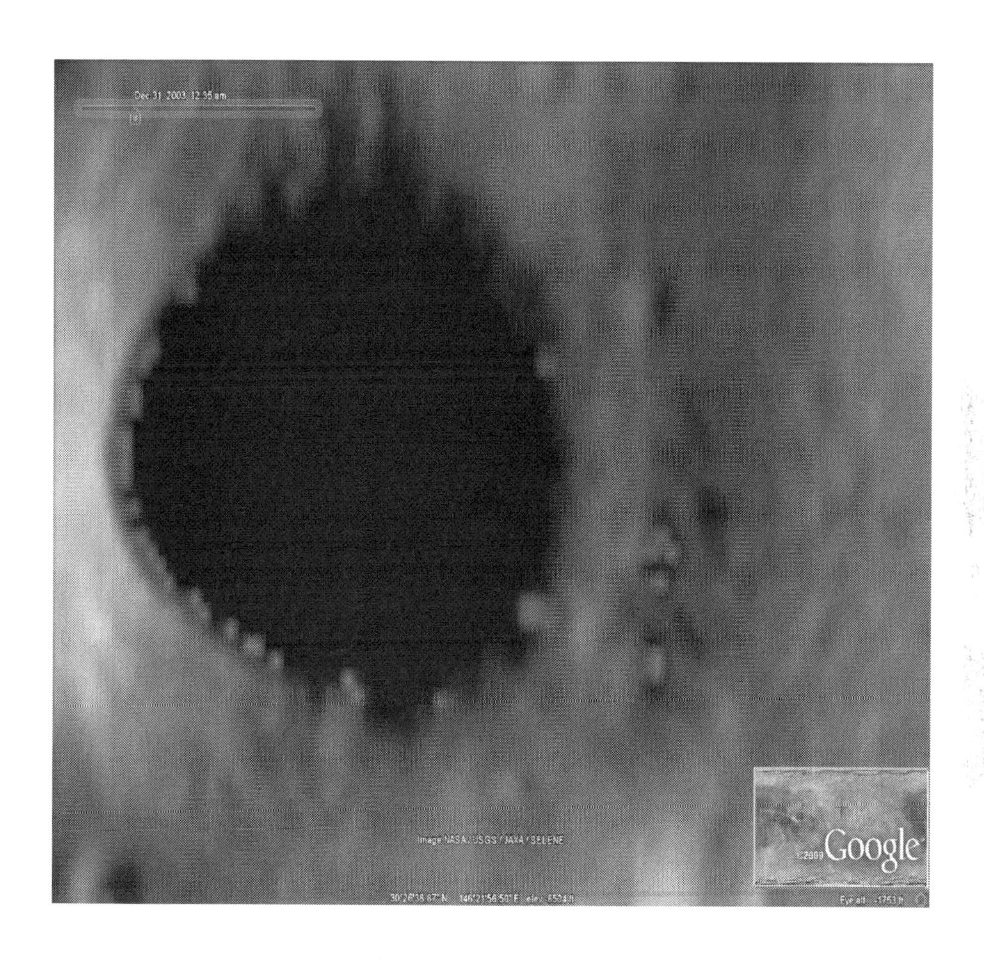

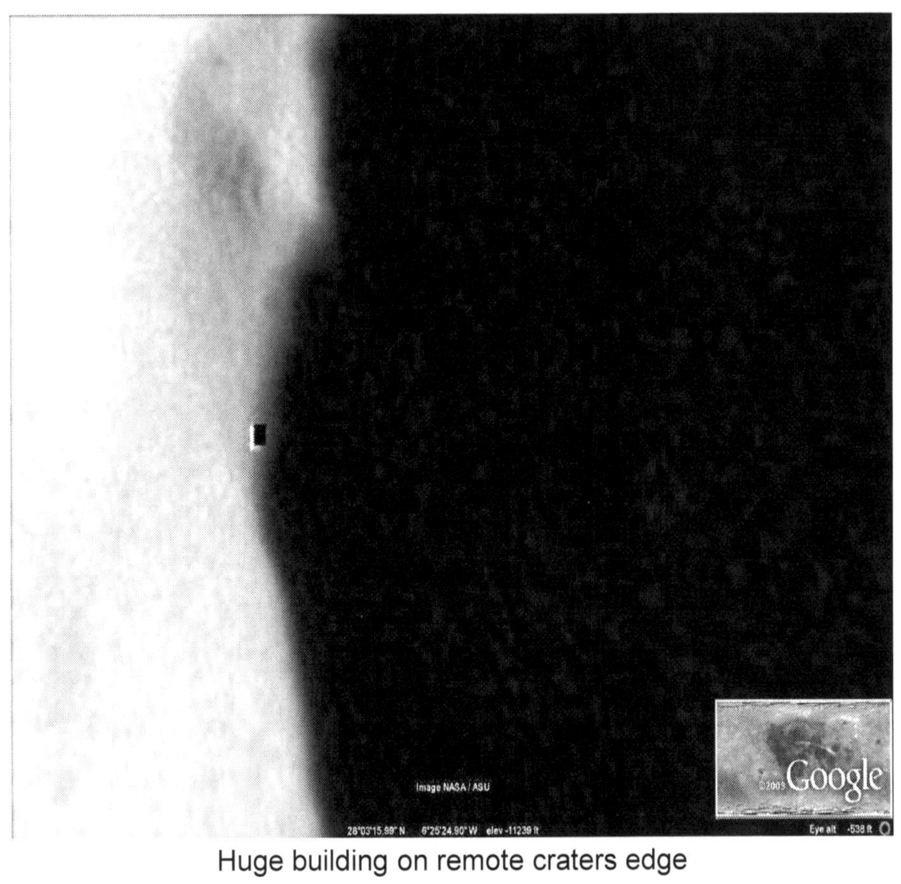

Huge building on remote craters edge

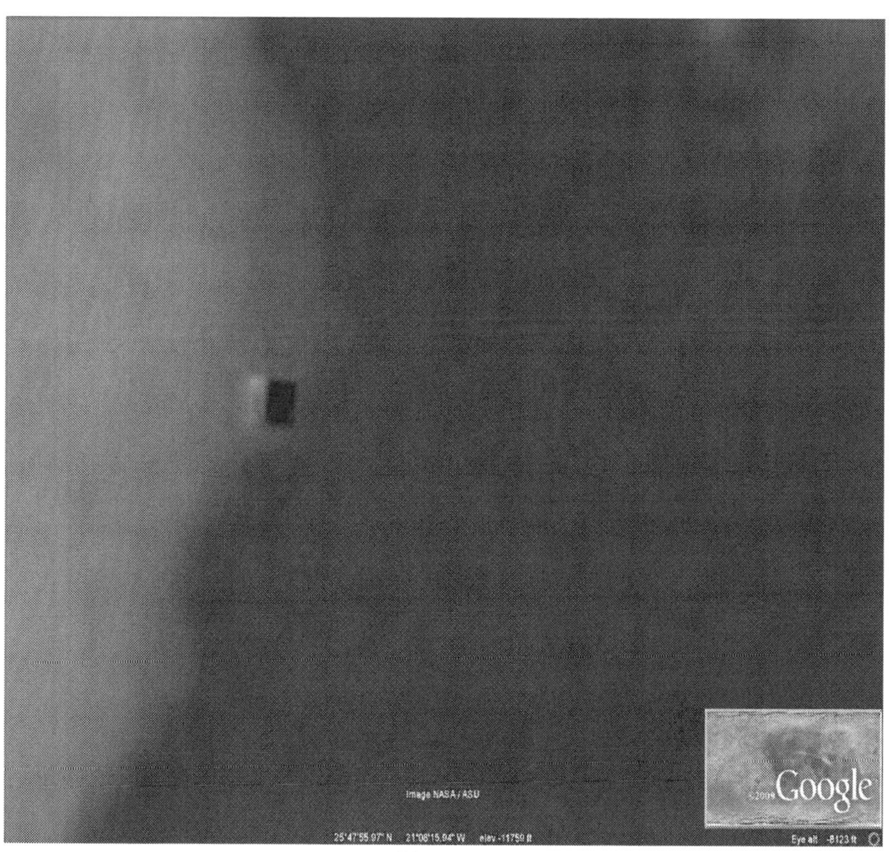
Image NASA / ASU
Google
25°47'55.97" N 21°08'15.94" W elev -11759 ft
Eye alt -8123 ft

Closeup

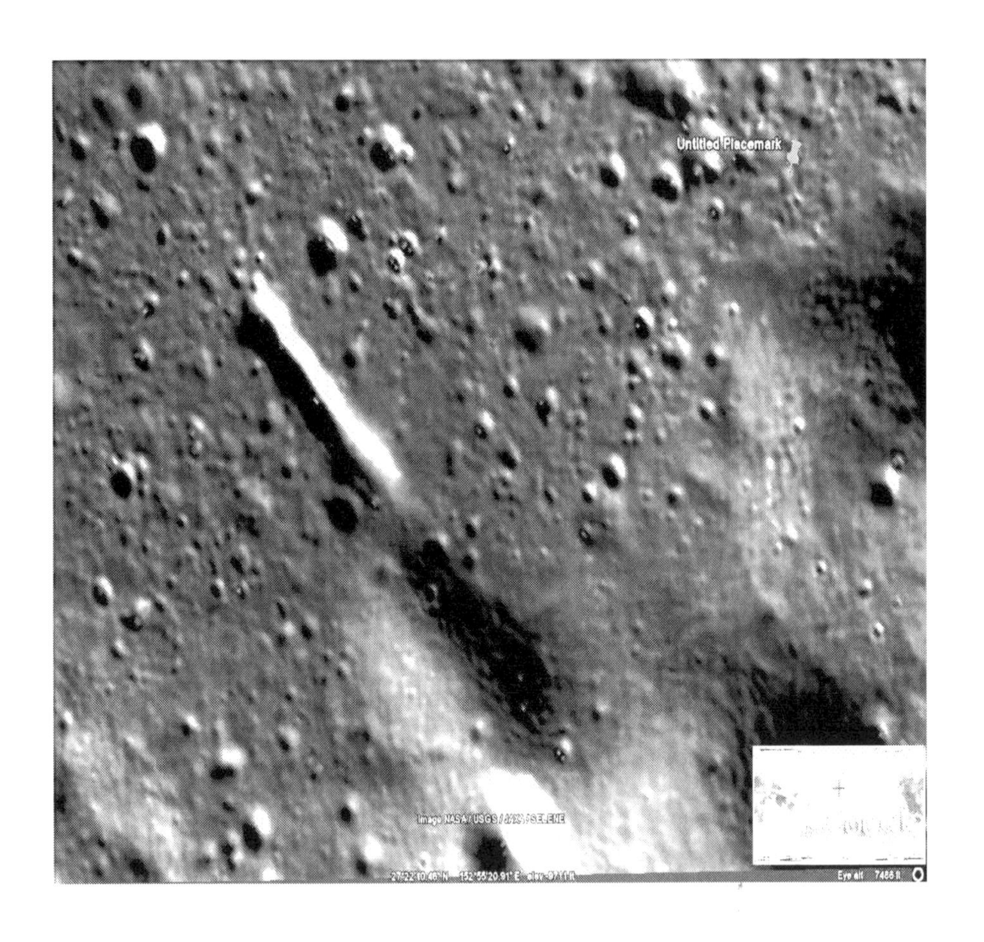
Untitled Placemark
Image NASA / USGS / JAXA / SELENE
27°22'10.46" N 152°55'20.91" E
Eye alt

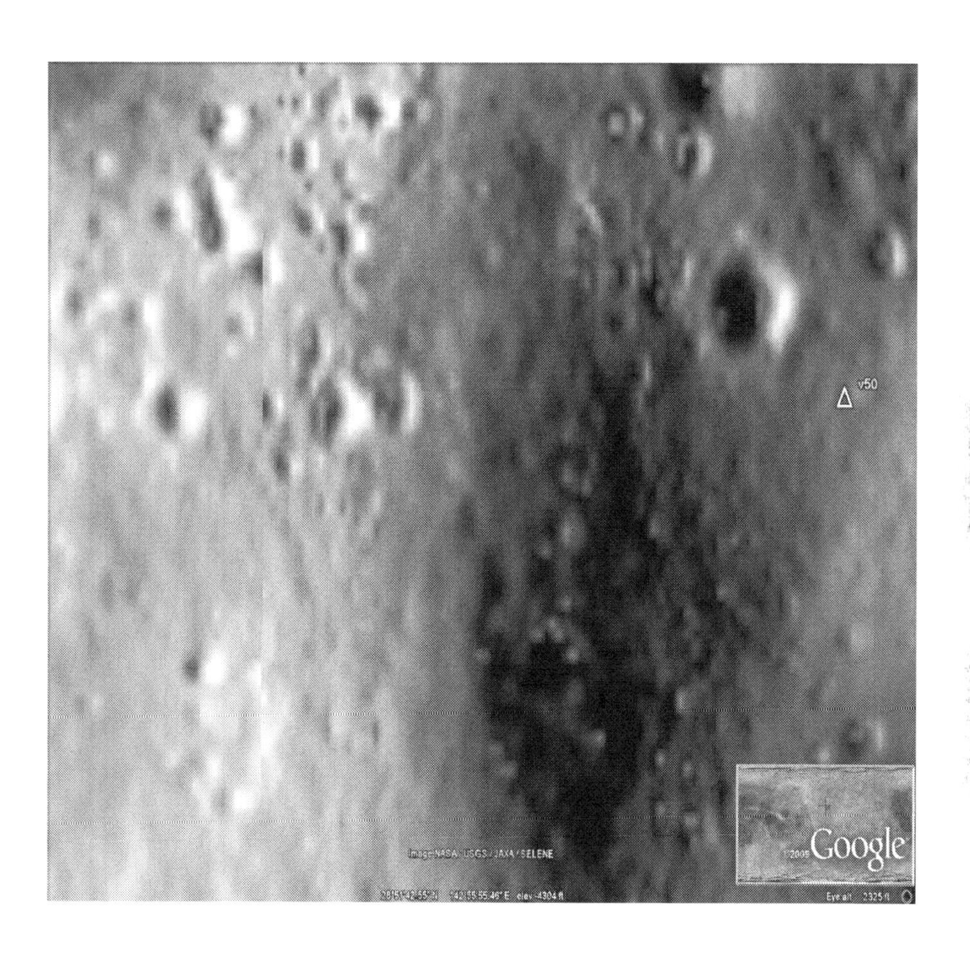

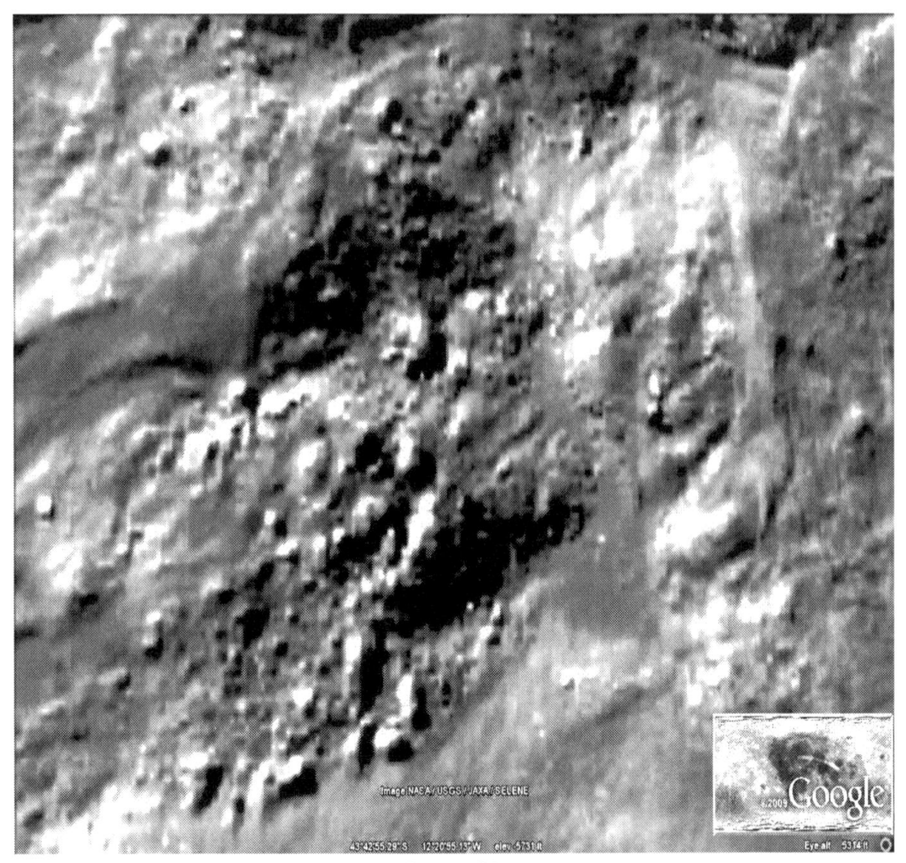
Image NASA / USGS / JAXA / SELENE
©2009 Google
43°42'55.29" S 12°20'55.13" W elev -5731 ft
Eye alt 5314 ft

Moon City

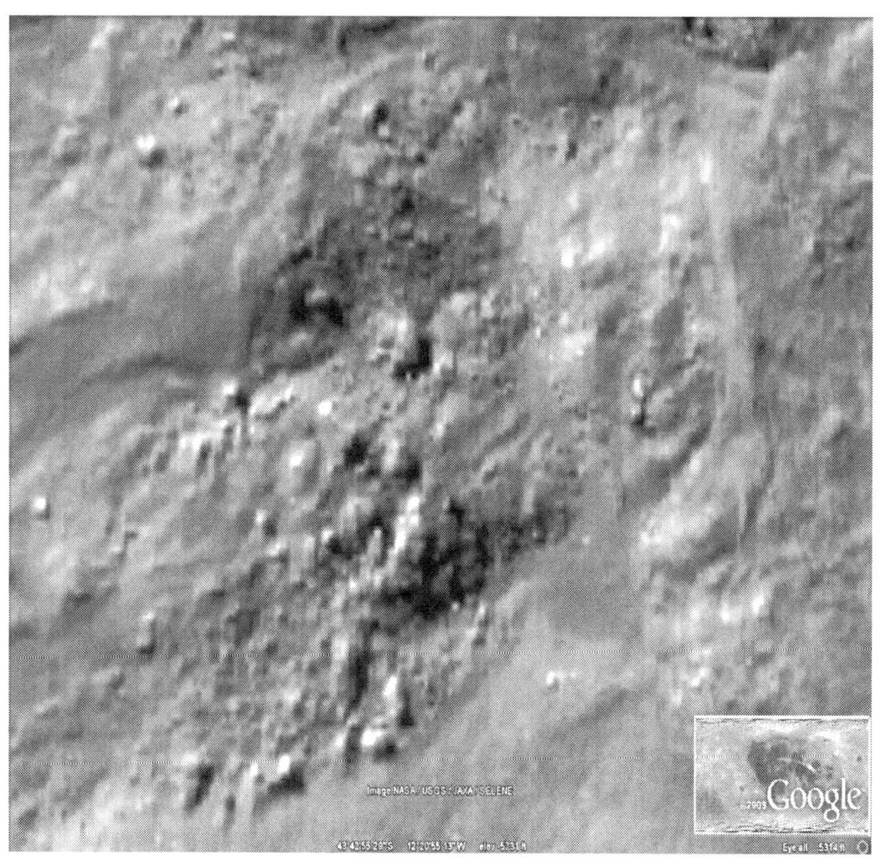

Closeup

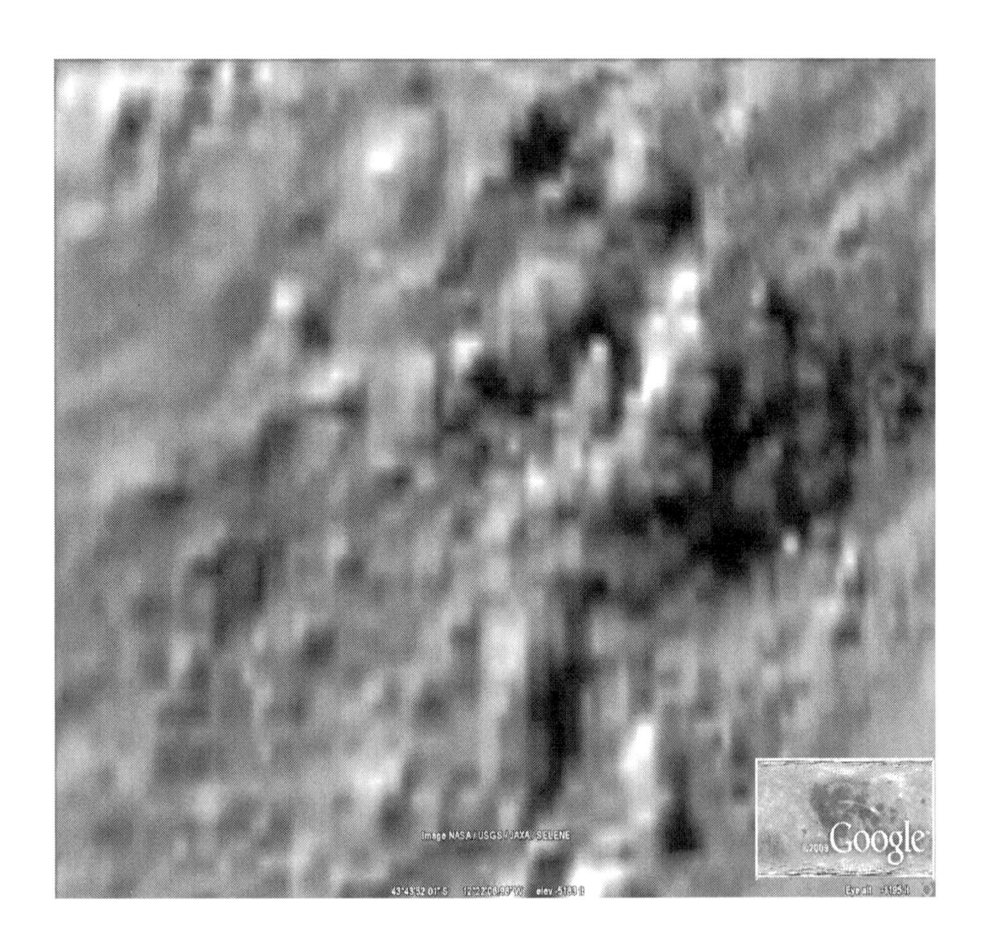
Image NASA / USGS / JAXA / SELENE
©2009 Google

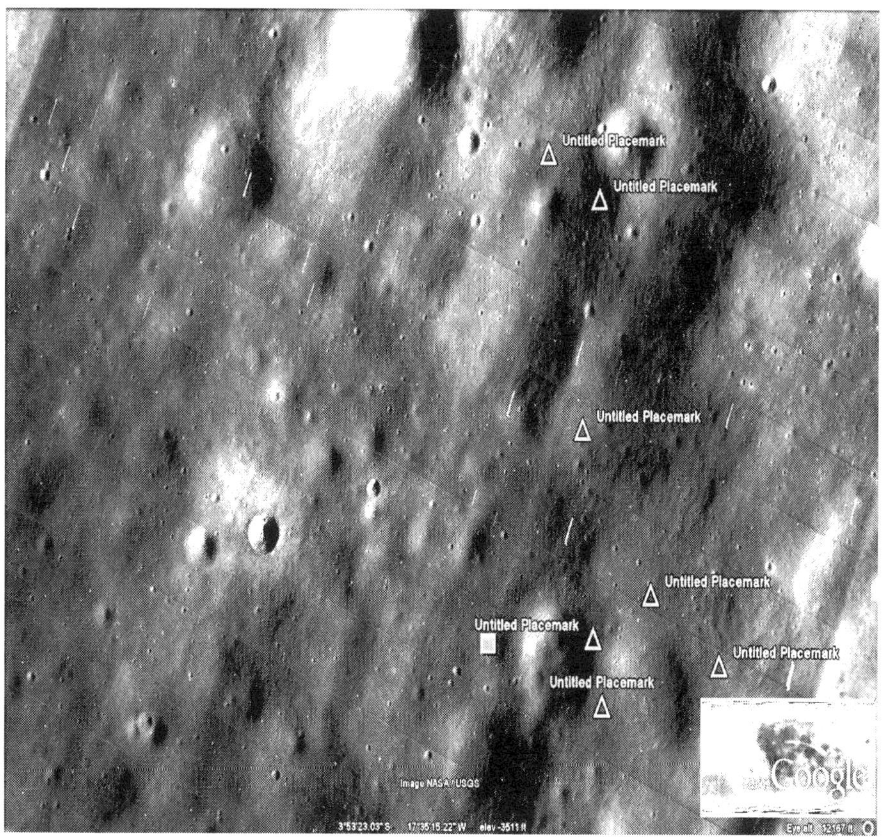

Field of coils

Dec 30, 2003 10:56 pm
Image NASA / USGS / JAXA / SELENE
Google
29°56'51.53" N 144°48'39.62"E elev -5066 ft
Eye alt 13170 ft

C shaped building

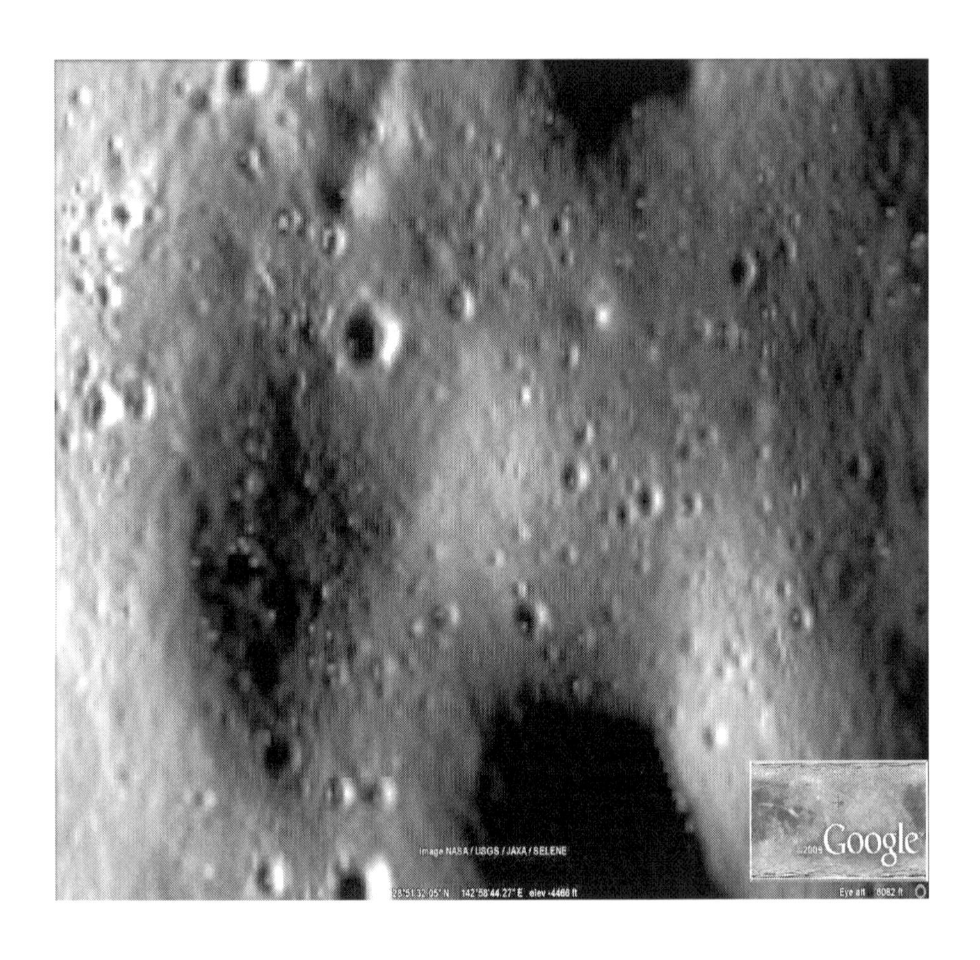
Image NASA / USGS / JAXA / SELENE
©2009 Google
28°51'32.05" N 142°58'44.27" E elev -4466 ft
Eye alt 6082 ft

City

Image NASA / USGS
©2009 Google
3°54'16.23" S 17°33'26.49" W elev -3436 ft
Eye alt

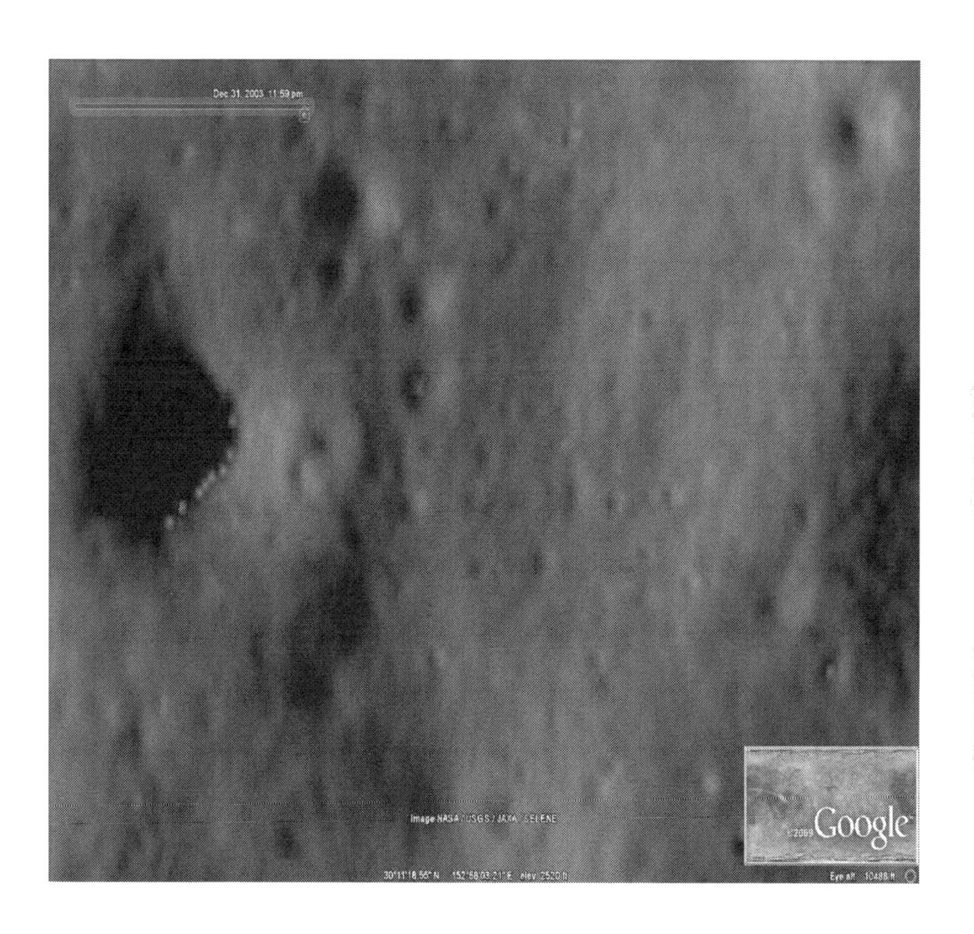
Dec 31, 2003 11:59 pm
Image NASA / USGS / JAXA / SELENE
©2009 Google
Eye alt 10488 ft

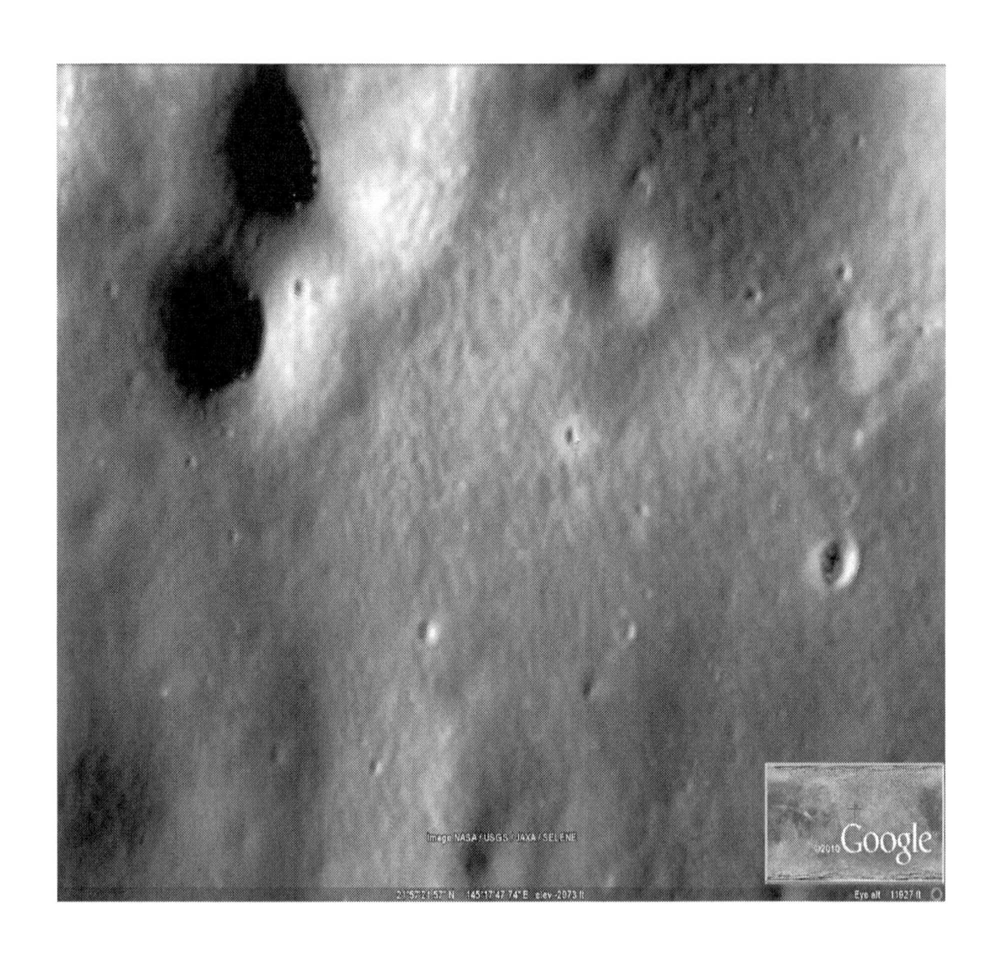
Image NASA / USGS / JAXA / SELENE
©2010 Google
elev -2973 ft
Eye alt 11927 ft

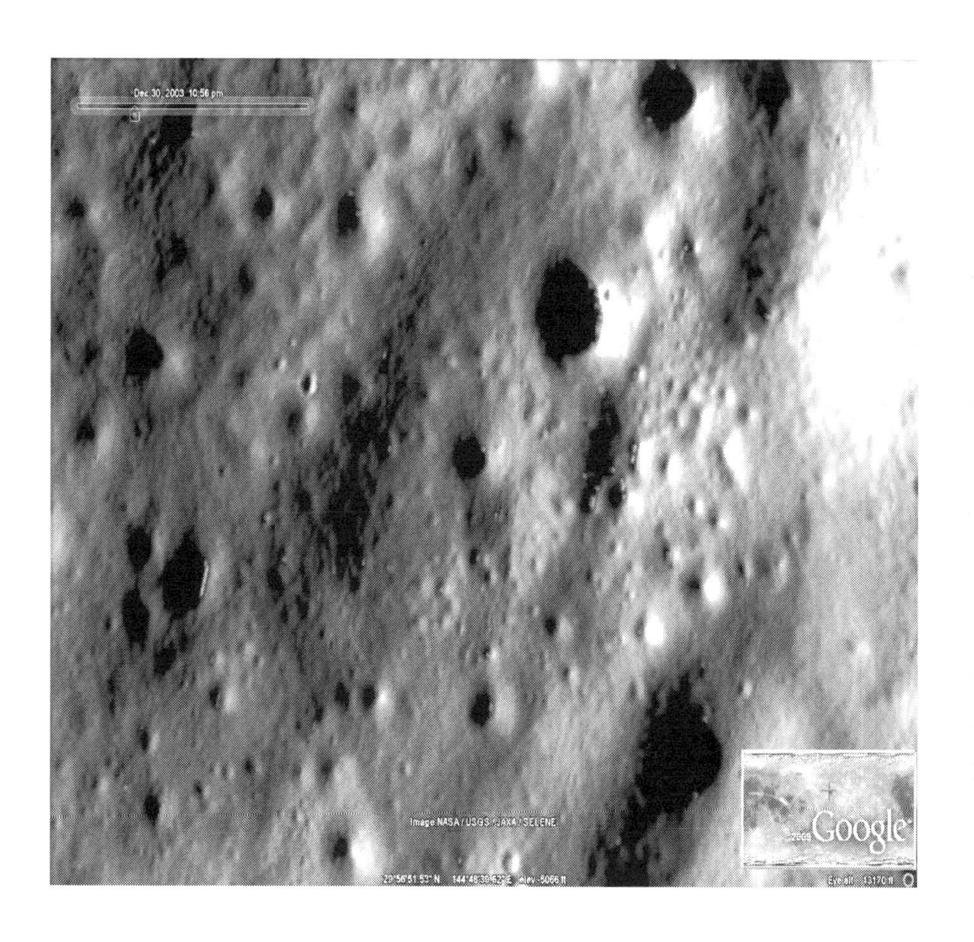
Dec 30, 2003 10:56 pm
Image NASA / USGS / JAXA / SELENE
Google
29°56'51.53" N 144°48'39.62" E elev -5066 ft
Eye alt 13170 ft

Closer view of city

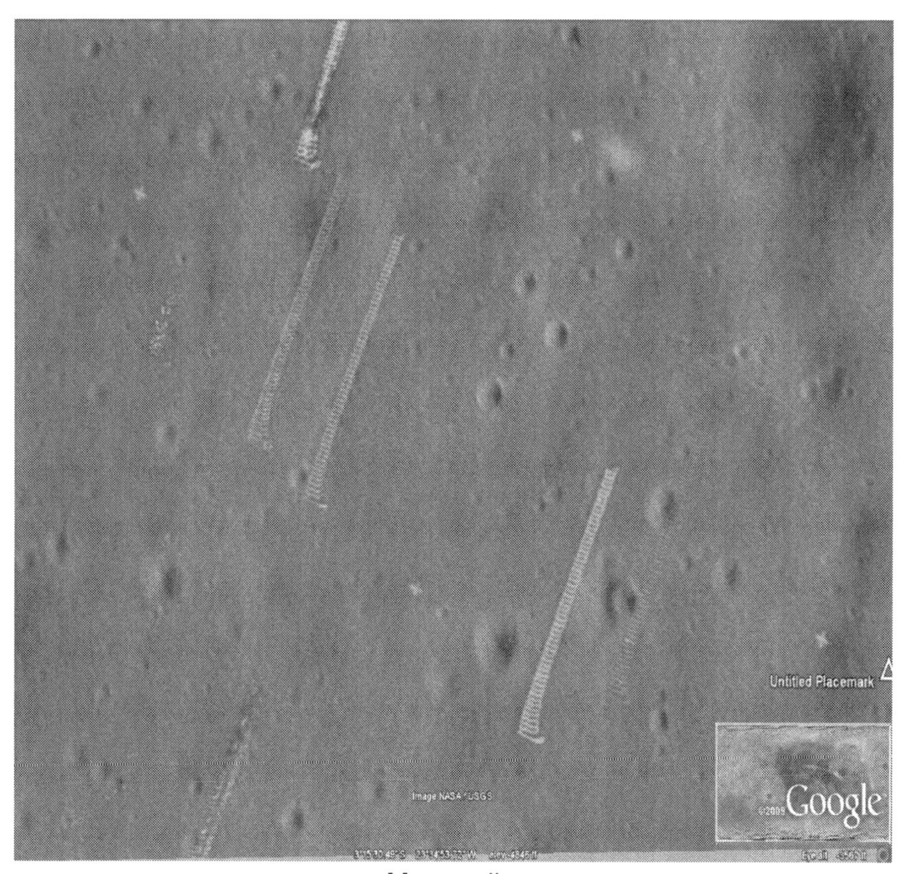

More coils

ABOUT THE AUTHOR

Kenneth Clark lives in the Tampa Bay area and is still involved in UFO research and investigation. He is currently working on new book about UFOs on Mars.

Made in the USA
Lexington, KY
24 February 2013